Practical Color Measurement

WILEY SERIES IN PURE AND APPLIED OPTICS

Founded by Stanley S. Ballard, University of Florida

EDITOR: Joseph W. Goodman, Stanford University

Practical Color Measurement

A Primer for the Beginner
A Reminder for the Expert

ANNI BERGER-SCHUNN

Translated from the German by the Author with the assistance of
MAX SALTZMAN

A Wiley-Interscience Publication

John Wiley & Sons, Inc.

New York / Chichester / Brisbane / Toronto / Singapore

Wiley Series in Pure and Applied Optics

The Wiley Series in Pure and Applied Optics publishes outstanding books in the field of optics. The nature of these books may be basic ("pure" optics) or practical ("applied" optics). The books are directed towards one or more of the following audiences: researchers in universities, government, or industrial laboratories; practitioners of optics in industry; or graduate-level courses in universities. The emphasis is on the quality of the book and its importance to the discipline of optics.

Copyright © 1994 by Muster-Schmidt Verlag

Authorized translation of the German edition published
by John Wiley & Sons, Inc.

Library of Congress Cataloging in Publication Data:

Berger-Schunn, Anni, 1928–
    Practical color measurement: a primer for the beginner,
a reminder for the expert/Anni Berger-Schunn; translated from the
German by the author with the assistance of Max Saltzman.
        p.   cm.
    Includes bibliographical references.
    ISBN 0-471-00417-0 (alk. paper)
    1. Colorimetry.   I. Title.
QC495.B388   1994
535.6′028′7—dc20   93-45347

Printed In the United States of America

10 9 8 7 6 5 4 3 2 1

Contents

Preface   vii

Introduction   ix

1. Description of Perceived Colors with the Aid of Numbers   **1**

    1.1. Light   3
    1.2. Sample   13
    1.3. Observer   19
    1.4. Calculation of the Tristimulus Values   26

2. Calculation of Color Difference   **35**

3. Metamerism   **57**

4. Color Measurement Systems; Measurement of Fluorescent Samples and Whiteness   **73**

    4.1. Spectrophotometers   74
        4.1.1. Light Sources   75
        4.1.2. Splitting the Light into Its Component Wavelengths   76
        4.1.3. Measuring Geometry   77
        4.1.4. Detectors   81
        4.1.5. Calibration of Instruments   81
        4.1.6. Accuracy of Instruments   84
    4.2. Colorimeters (Three-Filter Color Measuring Instruments)   86
    4.3. Influence of Instrument Variables on Accuracy of Color Difference Measurement   87
    4.4. Measurement of Fluorescent Samples   91
        4.4.1. Whiteness   97

5. Correlation between Reflectance (Transmittance) and Colorant Concentration, Examination of Colorant Strength, and Computer Color Matching   **99**

    5.1. Lambert–Beer Law   109
    5.2. Kubelka–Munk Equations   114
        5.2.1. Color Strength, Residual Color Difference   118

5.3. Computer Color Matching   127
      5.3.1. Determination of the Calibration Constants   130
      5.3.2. Calculation of the First Formula (Initial Match)   133
      5.3.3. Correction of the Formulas   143

**6. Influence of the Sample on the Accuracy of Color
   Measurements**   **146**

Appendix   **160**

Symbols and Terms   160
Formulas   163

Bibliography   **172**

Preface

Color measurement is a young science, although much about the nature of colors can be found in the works of Goethe and Newton.

Since 1931 there have been international agreements as to how colors can be described by numbers. At that time color measurements were done by only a few scientists because the measurements and the calculations were circuitous and time-consuming. None of the resources that are taken for granted today were available at that time. In industry color measurement has been used since about 1940. Industry recognized early on the value of this technology. Wide application of color measurement came with the availability of simple digital computers. Digital computers combined with color measuring instruments provided color measuring systems that were simple to use. In a very short time reliable results became available. Today almost all of the industries that deal with colorants have color measuring systems. The systems are used mostly for computer colorant matching but also for quality control. The use of color measurement systems gives unquestionable advantages, but there are also disadvantages.

Among the advantages are the determination of the lowest cost of coloration and the much shorter time required for color matching. The possibilities for statistical quality control can be achieved only when data can be stored. This possibility will be more important in the future, because many companies ask for quality certificates and statistical production data, together with the delivered products, to guarantee the delivery of a product within a narrow tolerance range.

The disadvantages are not so easy to see. They arise from unsatisfactory measurement results. Such results are frustrating for the users of such systems and result in disputes between the seller and the buyer when both get different results from the same samples and then argue if the shipment is within the agreed-on tolerance. There are two reasons for this. First, the color sensation is an effect in the mind of the viewer. To describe color sensation with numbers or, better stated, to describe the color of a sample with numbers, special agreed-on premises are necessary, which are not always given. Second, the result of the measurement gives only the values of the measured sample, which seldom represent the color of the whole lot.

While the small number of people who did color measurements in the early years had exact knowledge of the principles involved, the users today often blindly believe the printed numbers. They frequently have only little knowledge of the principles of color measurement and no knowledge of the limits of accuracy or precision of the printed results. The producers of color measurement systems help to make them uncertain; they sell fascinating software that blurs the limits. They also print the results with too many decimal places and confuse the users with an accuracy that is not there. The users of today sometimes do not detect obvious erroneous results of measurements.

This need not be, because there are many very good textbooks available. They describe both the principles of color measurement as well as the techniques of application. All of them seem to have the same disadvantage: they are too academic and too extensive. Therefore they are not carefully read. This book tries to describe the principles and particularly the limitations of color measurement. It is written as simply as possible with a great number of examples, which are admittedly sometimes not as simple as they should be. The reader shall decide whether this is a simple book.

I thought during the writing of this book that the readers would be employees of industries that deal with colors—dyehouses, factories for paints, printing inks, and plastics; paper mills and manufacturers of colorants; as well as managers, who have to interpret the results of color measurements. This book should also be useful for producers of finished products such as painted cars, molded plastic parts, and textiles. Producers of films, video TV tubes, and so on will not find their particular problems addressed in this book. The principles of color measurement are, however, valid for them, too.

In this book I make no claims to completeness, to the history and the development of color measurement. I describe only what every user of color measurements should really know to properly do the job. It is required that the reader of the book who does color measurements be provided with a computerized instrument. The computer then does the mathematically simple but often extensive calculations.

In this book the sources of the contents are not always given. The bibliography is small. I hope that the writers of the many good papers about color measurement understand the reason for this limitation. It is obvious that the theoretical background for this book comes from the published literature, and only a small part originates with me. The practical knowledge was obtained from many years of practical professional experience.

For the reasons discussed above not all figures have been newly drawn. They are taken from other publications with permission of the publishers. The sources of the figures are given with each of these figures.

The samples discussed in this book have been given to me by colleagues in different industries. Also the color measurements have been done by them. In most cases the samples are taken from normal production; a few were specially prepared.

My special gratitude belongs to Professor Max Saltzman. Without his persuasiveness this book never would have been written. I also thank Dr. Andreas Brockes, my colleague of many years, for reading this book both critically and constructively.

ANNI BERGER-SCHUNN

Introduction

As stated in the Preface, this book is different from all the very good textbooks available that deal with color measurement. It is different from all the other textbooks in that the coverage of the subject is much less extensive. Only what every user of color measurement systems really needs to know is covered. Many examples help to understand and to critically interpret the numbers that are given by color measurement instruments.

Mathematics cannot be wholly dispensed with. As far as possible the formulas used in color measurement are not derived. I only describe what underlying assumptions and calculations lead to the formulas. In order not to interrupt the reader's train of thought, all long formulas are given in short form only.

In the Appendix, the symbols used are put together with short explanations. All important formulas for color measurement are also repeated in the same Appendix in the order they are described in the text.

As already stated in the Preface, I do not attempt to furnish a complete bibliography of all the papers that deal with color measurement or related fields. Apart from textbooks and reference books, I cite only those papers that are mentioned in this book. Because of the rapid development of the techniques of color measurement, the bibliography lists only books published in the last 15 years.* Not all books that were written in this time are listed.

If this book stimulates a desire for more information, the readers can broaden their knowledge by reading the many good available textbooks about color measurement. Most of them have a detailed bibliography.

The content of this book essentially deals with the problem of getting numbers that describe colors. With the help of these numbers answers can be given to the questions:

"How large is the color difference between two samples to be matched?"

"How can I reproduce the color of a sample with the help of the measuring technique?"

In brief, the answers to these two questions are known as *quality control* or *quality assurance* and *computer color matching*. I emphasize the importance of understanding the limitations of the color measuring technique. This helps to explain why the results are not always satisfactory.

---

*See textbooks listed under Section 1 of the Bibliography, at end of the book following the Appendix.

The book has some paragraphs set in smaller type. They are either clarifications to the text above, or thoughts that are connected with the text. These paragraphs can be omitted without missing important thoughts in the text.

In order not to interrupt the reader's train of thought, all long formulas are found in an appendix (p. 163). All important formulas for color measurement are repeated in the same appendix.

The reader will see in the text some tables that are very large; one is in Chapter 2 (p. 49) and several are in Chapter 5 (pp. 122, 124, and 132). The reader should not be disturbed by these. It is not necessary to read all the numbers though they provide much information. It is worthwhile to examine the numbers that are discussed in the text.

The vocabulary used in this book is the internationally accepted nomenclature of the CIE as far as it is useful (CIE = Commission Internationale de l'Eclairage; International Commission on Illumination). It compiles (and distributes) information recommended for standards that are generally adopted by National Committees (see Bibliography). For better clarity, sometimes other words are used. Where this is done, the term is indicated the first time (initially) and the standard word is also given.

Because the book is not large and the contents are well organized, no subject index is included.

ANNI BERGER-SCHUNN

Practical Color Measurement

1

Description of Perceived Colors with the Aid of Numbers

The visual perception of colors is one of the five senses. It is remarkable that people have been successful in describing the color of samples, which are perceived by the eye and processed in the brain, with numbers.

For colors to be seen, light is necessary: colors cannot be seen in the dark. In addition, the color of incident light has a large influence on the perceived color. A familiar example of this is the spotlights used in the theater. If they change color, the objects on the stage, for example, the dresses of the dancers, also change color. The same dresses can show many different colors as the light changes. Everyone knows how much the complexion can change with a change in the kind of illumination. Especially noticeable is the pale complexion seen when looking in a mirror in a room illuminated with special fluorescent lamps. On streets illuminated with sodium vapor or mercury vapor lamps the skin and the color of other familiar objects have no similarity to the memory colors. Fabrics produced with different materials may have the same color for all materials when illuminated with one light source. By illuminating them with another light the different materials can have different colors. Conversely, a white tablecloth appears white to the human eye generally independent of whether the table is illuminated with daylight or with yellowish candle light. A glossy object changes color as it is illuminated and matched in different directions.

The perceived color of an object, which is influenced by the color of the light illuminating it, is determined by color measurement. We see in our surroundings objects with many different colors, although all the colors are illuminated by the same light. It is amazing how many different green colors can be seen at the same time outdoors. The reason for this is the differential absorption of the incident light by the objects. With color measurement we determine the part of the light that is not absorbed; that is, we measure how much light is reflected or transmitted from the sample to be tested.

The third aspect or component of color vision required to describe the sensory impression of color is the human eye or, better, the eye–brain combination. In spite of all the research and in spite of all the acquired knowledge it is not entirely clear how to unequivocally describe the power of vision. It is certain that the eye has receptors for light, which transform the light falling on it into stimuli, which are sent by nerves to the brain, where the color perception originates. Color perception is different for all people with normal vision (around 96% of the population have "normal vision," the remaining 4% have abnormal color vision—see Section 1.3). It changes also when people get older. In daily life this is not specially conspicuous and therefore also not generally noticed. When matching pairs of samples, it happens often that the color difference between the two samples is judged differently by different observers. This is true not only for the size of the difference but also for the nature of the color difference. People who do color matching find or encounter samples, mostly olive or red-brown ones, which, although uniform, have a spot with another color. This spot moves on the sample when different parts of the sample are matched (moving of the head). This is an indication that the receptors in the eye are not uniformly distributed and that the color perception depends on the part of the eye on which light from the object is fixed.

Finally it must be pointed out that color perception depends on the color of the background of the sample. Also structure, gloss, and metallic effects (such as seen in modern car paints) influence the perceived color. Therefore it is difficult to match two samples that differ in structure—velvet and silk, broadcloth and corduroy, or leather and fabric. Visual judgment as well as measurement depends on how the two samples are arranged and how the light falls on the samples. This is true also when a glossy and matte sample are to be matched. Such differences in appearance influence the description of colors with numbers.

In practice nobody demands that the color sensation be described "absolutely true" with numbers. It is sufficient to find a method which tells us if two samples look alike, because this is the only question that is asked in daily practice. Whether it is asked if the color of the production is equal to the color of the standard, or whether the match is equal to the submitted sample, or whether the color of a road sign has the values given in the standard. The last question is still more difficult to answer, as will be discussed in Chapter 4 (p. 85). It is therefore enough to determine values that are equal when two samples look alike. In the beginning color measurement should answer only this question. The so-called *basic colorimetry* cannot answer other questions.

Unfortunately such pairs of samples are very seldom encountered in practice. The samples that are matched in practice are seldom absolutely equal in color. Our perception and the measured values say that both samples, if at all, match under only one light source and for one observer. Usually two samples that should be equal have a more or less large color difference. To describe that color difference with numbers, we must use the so-called *advanced colorimetry*; its statements are neither absolute nor definite. Nevertheless, agreements on how to calculate color differences have worked in practice.

Of the three factors that determine the perception of color, two—the incident light and the change in the light caused by the object or the sample—can be determined with the help of physical measurements. The sensation of the observer can be determined only indirectly. The three quantities and their inter-relationships are described below.

1.1. LIGHT

Light is part of the electromagnetic spectrum. This is part of a continuous band of radiation that includes X-rays; ultraviolet (UV), visible, infrared (IR), or thermal radiation; and radio and television waves. All these kinds of radiation differ only in the number of oscillations in a certain unit of length. For color measurement the wavelength of one oscillation is chosen as the measuring unit. It is expressed in the internationally agreed-on unit *meter* (m).

To use electromagnetic radiation technically, we need receivers that are sensitive to the wavelength range of interest. Radio waves are useless as long as there are no radios to receive them. The part of the electromagnetic radiation to which the eye is sensitive is called *visible light* or simply *light*. The human eye is sensitive to light as electromagnetic radiation with wavelengths between about 380 and 780×10^{-9} m [1×10^{-9} m $= 1$ nm $= 1$ nanometer (nm)]. For color measurements it is often restricted to 400–700 nm. Radiation with short wavelengths is recognized as blue. Changing to longer wavelengths, the color sensation becomes green, yellow, orange, or red. The radiation of all natural light sources as daylight or fire contains all wavelengths of the visible region of wavelengths (of the visible spectrum). That is true also for most artificial light sources such as incandescent lamps or fluorescent lamps.

The amount of radiation (the radiation power) varies from wavelength to wavelength for different light sources. They emit light of varying brightness and hue—bright blue daylight, yellowish incandescent light, and so on. There are methods for measuring the radiation power at each wavelength. For color measurement we need not be interested in such methods. We can assume that the spectral power distribution for each light source of interest for color measurement is known. (An important exception is the need to know the spectral power distribution of the illuminating light when measuring fluorescent samples, as will be discussed in Chapter 4.)

In general, color perception is independent of the amount of light that enters the eye. The eye is so constructed that the amount of light that enters it is controlled, to a great extent, by a diaphragm, the iris. If the amount of light is too small, we can see no colors. This everybody has recognized during a nighttime walk. If the amount of light is too large, we are blinded and cannot recognize colors.

This remark is also valid for instruments (Chapter 4). If the radiation that goes to the detector is too small, its sensitivity is not large enough to measure the radiation exactly and reproducibly. If the amount of radiation is too large, its accuracy may be affected. The latter case is very seldom encountered in practice.

For color measurement it is therefore agreed, to describe the radiant power of light sources only relative to the radiation of light at a single wavelength. Radiation with the wavelength of 560 nm is often given the value 100. The relative spectral distribution of light sources is indicated by the letter $S(\lambda)$. The letter in the parentheses indicates that the function S is dependent on the wavelength λ. The relative spectral power distribution is given in the diagrams (see Figure 1.1-1 and following), where the relative radiant power (the relative energy) (vertical axis = ordinate) is shown as a function of wavelength (horizontal axis = abscissa). As previously stated, more than once, for the perception of colors only the wavelength range between 400 and 700 nm is of interest. Daylight, however, contains radiation with wavelengths from 300 nm on. The wavelength range from 300 to 400 nm plays no direct role for color vision. Indirectly it has, as will later be shown, a remarkable influence. The relative spectral power distribution of daylight and light sources with relative spectral power distributions that are similar to those for daylight are therefore often shown from 300 to 700 nm.

The spectral power distributions for several light sources are discussed below.

We start with the spectral power distribution of the black body. It has no technical importance, but its spectral power distribution is referred to often in the description of other light sources. The black body is a hollow structure, usually a sphere, whose interior is black. It has a small opening. If we heat up the black body, we get radiation from the opening. The color of the radiation changes with the temperature of the black body. At low temperatures the opening looks dark red. The higher the temperature, the yellower (and lighter) the radiation. Figure 1.1-1 shows the relative spectral power distribution of the black body—from here on referred to as *spectral power distribution*. The measure for the temperature is the absolute temperature K. The symbol (K) for Kelvin is used without the word *degree* or the symbol °. Absolute temperatures are equal to degrees Celsius (°C; centigrade) plus 273. We can see that the curves change systematically. With increasing temperature more and more radiation with short wavelengths is emitted.

The spectral power distribution of the older artificial light sources—kerosene lamp, candle, incandescent lamp—is similar to that of a black body, because they are all thermal radiators. Figure 1.1-2 shows the spectral power distribution of standard illuminant A. (Standard illuminant A today is standardized as the spectral power distribution of a black body, a Planckian radiator. In the original recommendation the relative power distribution of a tungsten-filament lamp was standardized as standard illuminant A. Both are very similar.) The spectral power distribution of the incandescent lamp is assigned to the temperature of the black body. It is called *color temperature*. The color temperature corresponds to the spectral power distribution of the black body, which is most similar to that of the incandescent lamp. The spectral power distribution with a color temperature of 2856K is standardized and is called *standard illuminant A.*

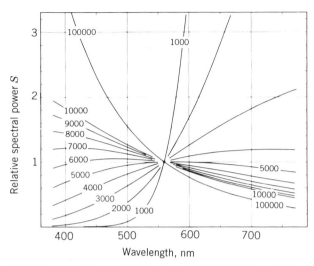

Figure 1.1-1. Relative spectral power distribution of the black body at different temperatures. The curves are standardized so that the spectral power is 1 for all temperatures at 560 nm (from Wyszecki and Stiles[1]).

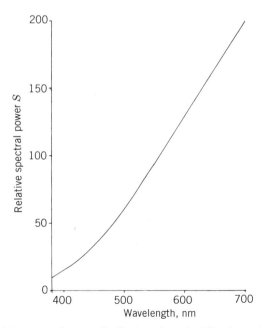

Figure 1.1-2. Relative spectral power distribution of standard illuminant A (black body with a color temperature of 2856K). The spectral power distribution is similar to that of an incandescent lamp.

To describe a light source with fewer words, the classification with a color temperature is done also when the curves of the spectral power distribution deviate more or less strongly from that of the black body. Such a color temperature is called *correlated color temperature*.

Figure 1.1-3 shows the spectral power distribution of daylight. The values that are established by the CIE are shown. These curves are also marked with the color temperature. It can be seen how much the color temperature and therefore also the spectral power distribution can differ. Further, it shows that the curves of daylight are at least similar to that of the black body. That the color of daylight can differ strongly is, on the other hand, generally known. Direct sunlight looks yellowish and therefore has a low color temperature. On a cloudy gray day the light, as the name indicates, is neither yellowish nor bluish. Such daylight has a color temperature of about 6000K. Light that comes from a clear blue sky may have a color temperature as high as 30,000K.

Because the color temperature (e.g., the spectral power distribution of daylight) changes from minute to minute, from day to day, from place to place, internationally a mean daylight has been standardized. For standardization the spectral power distribution of 6500K was chosen. Daylight with the standardized color temperature of 6500K is called *D65* (the standardized symbol is D_{65}). In Figure 1.1-4 the spectral power distribution of standard illuminant D65 is shown again by itself. Also standardized are the standard illuminants D50, D55, and D75, which should be used only in special cases.

Because the relative spectral power distribution in colorimetry is, as will be shown later (p. 26), generally used only for calculation, it does not have to exist as a light source. Therefore the standardized relative spectral power distribution is called a *standard illuminant* and not a *standard source*. Because colors are not only measured but also matched visually and because in special cases the light sources in the measuring instrument should have a spectral power

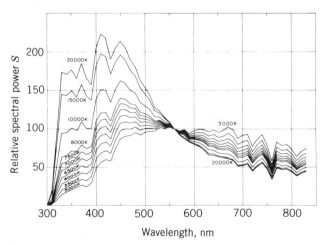

Figure 1.1-3. Relative spectral power distribution of daylight (from Wyszecki and Stiles[1]).

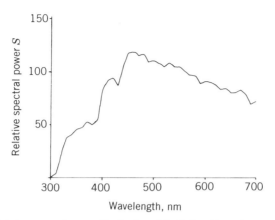

Figure 1.1-4. Relative spectral power distribution of daylight with a color temperature of 6500K. The spectral power distribution is called *standard illuminant D65*.

distribution that agrees with a standard illuminant, we try to get light sources whose spectral power distribution is equal to that of a standard illuminant.

Before standard illuminant D65 was recommended, after a large number of measurements of the spectral power distribution of daylight at different places had been made, standard illuminant C was the illuminant used as average daylight. The color temperatures of standard illuminant C and D65 are similar; however, both illuminants differ considerably from each other. Standard light source C was produced by filtering a tungsten-filament lamp with a standardized color temperature. The spectral power distribution of the light was measured and standardized as standard illuminant C. Standard illuminant C corresponds only in the visible wavelength range to the spectral power distribution of daylight, because it is difficult to filter tungsten-filament light in such a manner that the filtered light has a spectral power distribution that corresponds to daylight in the UV region. In this wavelength range the energy is much too low as compared with daylight. Standard illuminant C is, therefore, along with standard illuminant D65, only a standard illuminant and not a practical standard source. Standard illuminant C should no longer be used in color measurement.

Artificial daylight is needed for illuminating rooms and also as a light source for visual color matching. Artificial daylight lamps therefore have an important industrial and scientific significance.

The most important daylight lamps are fluorescent lamps. Xenon lamps are important technically for both visual matching and as a light source in measuring instruments. Both types of lamps are, in contrast to the lamps described above, not thermal radiators but gas-discharge tubes.

If we discharge rare gases as neon, argon, or xenon but also metallic vapors as mercury vapor or sodium vapor with the help of electricity, they also emit light. Contrary to thermal radiators, such light contains only the few wavelengths that are characteristic of the excited molecules in them. When we increase the pressure inside the tubes, because of the interaction of the molecules, a continuum of radiation is also emitted. Discharge lamps with a low internal pressure are used mostly for calibration. Discharge lamps used for illumination have a more or less high internal pressure and therefore also a continuum.

Figure 1.1-5 shows the spectral power distribution of a high-pressure sodium lamp, as used for the illumination of streets. The emitted light consists essentially of several yellow wavelengths. Figure 1.1-6 shows the spectral power distribution of a high-pressure mercury lamp, which is used also for the illumination of streets and of stadiums. In this lamp also a large part of the emitted light consists of the peaks of the mercury vapor.

A discharge lamp with relatively small peaks but a high amount of continuum is the xenon arc lamp. The spectral power distribution of that lamp is shown in Figure 1.1-7. At first glance we can see that the distribution is similar to that of standard illuminant D65. By filtering, this distribution can be adapted to that of daylight. At this time there is no better substitute for daylight. Figure 1.1-8 shows the spectral power distributions of standard illuminant D65 and of a filtered xenon arc lamp.

Today in many color measuring instruments the high-pressure xenon arc lamps described above are no longer used; instead xenon flash lamps are used. Their spectral power distribution is somewhat different from that of the high-pressure lamp. The ratio of line spectrum to continuum is higher in the flash lamp. Because of the lower internal pressure of xenon arc lamps, fewer safety precautions are necessary. Xenon flash lamps differ from high-pressure xenon lamps in terms of spectral power distribution as other gas-filled lamps with low pressure differ from those with high pressure. This means that the portion of the continuum is smaller and that of the lines is greater. Their spectral power

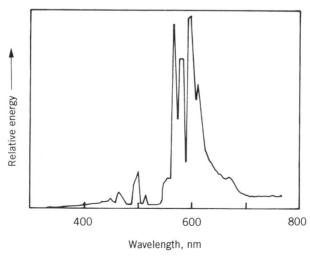

Figure 1.1-5. Relative power distribution of a high-pressure sodium lamp (from McDonald[1]).

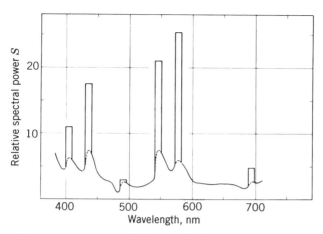

Figure 1.1-6. Relative spectral power distribution of a high-pressure mercury lamp (from Wyszecki and Stiles[1]).

distribution therefore does not correspond with daylight as well as that from high-pressure lamps. Nevertheless, they are successfully used in color measuring instruments, because they contain enough UV light.

As will be described in Section 1.2 (p. 13), part of the light that falls on a sample is absorbed. The absorbed light warms up the sample and is emitted from the sample as invisible thermal radiation. This event is taken for granted since many people have burned their fingers on a black car standing in the sun. This never happens with a white car. There are also materials that convert the absorbed light only partly into heat but emit the other part as radiation. Part of the absorbed light always is lost, and the emitted radiation always has lower energy than does the absorbed radiation. To state this phenomenon in terms of colorimetry, the emitted light always has longer wavelengths than the absorbed

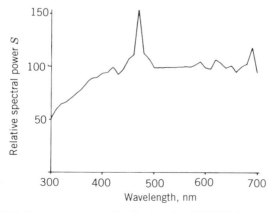

Figure 1.1-7. Relative spectral power distribution of a 150-W high-pressure xenon lamp.

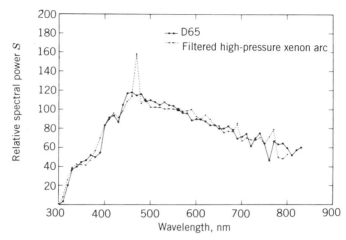

Figure 1.1-8. Relative spectral power distribution of filtered high-pressure xenon arc light in comparison to D65 (from Wyszecki[2]).

light. Depending on the time required for the emitted light (luminous light) to return to zero after switching off the illuminating source, we say the material is fluorescent (enduring for short time) or phosphorescent (enduring for long time). For the sake of simplicity we will speak here only of fluorescence.

For fluorescent lamps we use the phenomenon discussed above. Fluorescent lamps are mercury arc tubes with a low internal pressure. The inner wail of the tubes is painted with phosphors, which are activated by the radiation of the mercury vapor to fluoresce. Depending on the composition of the phosphors fluorescent lamps emit light that is similar to daylight or incandescent light. The color temperature of fluorescent lamps can vary in the range of 3000–6500K. Because the distribution of the radiation differs strongly from that of a black body as will be shown below, we only speak of their correlated color temperatures.

The spectral power distribution adds the wavelength peaks from the mercury vapor to the continuum that is emitted from the phosphors. There are three main types of fluorescent lamps. The spectral power distributions of two of them differ mostly in the longer-wavelength spectrum. The spectral power distribution of one of each type is shown in Figure 1.1-9. Lamps that have more radiation in the longer-wavelength region often add to their name the term "de luxe."

In the third kind of fluorescent lamps radiation is mostly emitted in three regions of wavelengths. They are called prime-color lamps or (e.g., in Germany and the UK) three-band lamps (Figure 1.1-10). Prime-color lamps have, among other things, a particularly good luminous efficiency. Therefore they are used often for the lighting of stores. Because their spectral power distribution is much different from daylight, the matching of samples for this light is especially difficult.

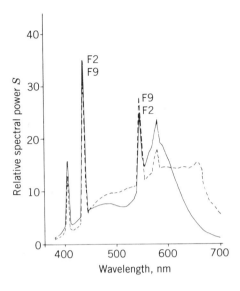

Figure 1.1-9. Relative spectral power distribution of two fluorescent lamps. Their spectral power distributions are recommended by the CIE. The correlated color temperatures of both lamps are about 4200K. Their color-rendering indices are 64 for F2 and 90 for F9.

The spectral power distribution of fluorescent lamps is not yet standardized today. The CIE has recommended, however, the temporary use of special spectral power distributions (F1 to F12).

How well colors are rendered by illumination with artificial daylight lamps in comparison to a thermal radiator or to daylight is characterized with a value called *color-rendering index*. Under a light source with a color-rendering index 100 all samples show the same color as under the reference lamp (thermal radiator or daylight) with the same color temperature. If the color rendering index is low, some colors, such as the color of the skin, look different. Light sources with a color-rendering index of at least 90 are called good. Fluorescent lamps have, depending on the type, color-rendering indices between 60 and 95, an important fact whose importance is often neglected.

There are more artificial light sources with other spectral power distributions. At the moment they are of no importance for color measurement and color matching.

Summary. The relative spectral power distribution of the illuminating light plays an important role in color measurement and for visual matching of pairs of samples.

In the results of color measurements the spectral power distribution of light generally is only a set of numbers used for calculation. An exception is the measurement of fluorescent samples.

The CIE has recommended the use of certain spectral power distributions as standard illuminants:

Standard illuminant D65 (6500K) for average daylight. Alternatively the use of D50 (5000K), D55 (5500K), and D75 (7500K) is also permitted.

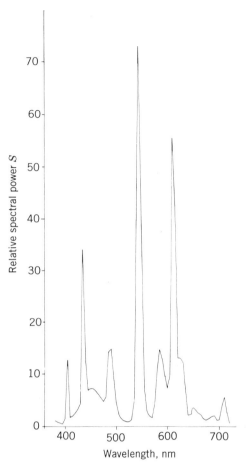

Figure 1.1-10.   Relative spectral power distribution of a prime-color lamp. Its spectral power distribution is recommended by the CIE (F11). The correlated color temperature is 4000K. The color-rendering index is 83.

Standard illuminant A (2856K, as incandescent light for illumination of rooms).

Illuminants F1–F12. These are the spectral power distributions of fluorescent lamps of different types. If fluorescent lamps are used, F2 (4230K), F7 (6500K), and F11 (4000K) are preferred.

The standard illuminants should be capable of being prepared as standard sources for visual color matching as well as for the measurement of fluorescent samples. This has been only partially successful for the standard illuminants D50–D75. Filtered xenon light is the closest approximation.

Light for visual matching therefore nearly never has the spectral power distribution of a standard illuminant. This is true for visual matching with daylight as well as for visual matching under so-called artificial daylight sources. As discussed in detail (p. 53), this is one of the reasons for the limitations of the significance of calculated tristimulus values.

1.2. SAMPLE

In the remaining text of this book all colored materials and objects are called "samples." This makes sense since, except in nature, all objects surrounding us are artificially colored. Matching the sketches of a designer, reproducing the submissions of a customer as well as the production samples that must match the production standard; all are samples to be tested. If natural products are tested or if they are patterns for color reproduction, they may be called "sample" as well.

As already stated in the introduction to this chapter, samples appear colored because they modify the incident light. The colored appearance is produced because the samples absorb a part of the light. The absorbed light is emitted generally as thermal radiation. The rest of the light is either transmitted or scattered. The scattered light also is partly reflected.

Samples can be divided into three groups:

1. *Transparent Samples.* Such samples absorb a part of the illuminating light. The other part goes unscattered through the sample. Transparent samples include eyeglass lenses, colored glasses, or transparent plastics of all kinds (e.g., drinking vessels; also color filters, which change the illuminating light in a specific manner). A well-known example of colored filters are sunglasses. A special kind of transparent samples are furniture lacquers, which can be colored or uncolored. Because transparent samples generally have a higher refractive index than air, the different refractive indices at the surface air–sample causes part of the light to be reflected. About 4% of the incident light is reflected. That means that clear transparent samples only transmit about 92% of the incident light. It is possible, as is often done with eyeglasses, to produce nonreflective surfaces.

2. *Translucent Samples.* Samples that not only absorb part of the incident light and transmit the other part but also scatter a part of the nonabsorbed light are called *translucent*. The scattered part is partly transmitted and partly reflected. Examples of translucent samples are lamp panels and lamp shades. It is difficult to visually match, measure, and reproduce translucent samples. Also translucent samples have the change in refractive index described above. Therefore they also reflect about 4% of the incident light at each surface.

3. *Opaque or Reflecting Samples.* Opaque samples either absorb the incident light or reflect it. No light is transmitted. Most of the colored objects that surround us are opaque. They can be painted walls, textiles, papers, or cars. If such products are not absolutely opaque, we make the samples opaque, where possible, by taking more than one layer for visual matching or for measurement.

It is possible to test if a sample is opaque, by placing the pile of samples on a black-and-white base. The contrast of the base should not be seen. We can examine it also by holding the pile against a bright light. The pile must have so many layers that no light is transmitted through the pile.

If we examine paints, the best thing is to draw them down on a black-and-white contrast base. The layer must be so thick that the base cannot be seen. If it is not possible to draw down directly one thick layer, we lay down several layers one above the other to get the necessary hiding. Plastic granules are tested by preparing samples of them, in the laboratory, which are thick enough to be opaque.

If the transparency of a sample is not a property to be tested, in industry mostly all of the samples to be tested can be made opaque using this technique. Therefore we will discuss mostly the measurement of opaque samples.

As stated above, samples look colored because the colorants in them (*colorant* is a collective term for dyestuffs and pigments popularized by Billmeyer and Saltzman) absorb part of the incident light. Colorants normally are artificially added, and a large number of the readers of this book will have to deal directly or indirectly with the coloration of samples. All raw materials contain colored components as impurities. This will be discussed later on (p. 121).

Depending on the kind and the amount of the colorants used, the different wavelengths of the incident light are differentially absorbed. The radiation that is not absorbed is reflected. It is seen by the eye, when looking at the sample, as color. The reflectance of a sample can be measured. The same is true for the transmittance of a nonturbid sample. What is said about reflecting samples is true for transmitting, nonturbid samples. If we measure the percentage of the light reflected at each wavelength, and if we plot the measured values as a graph, we get the reflectance curve.

Reflectance here is used as a collective term. The CIE recommends, depending on the conditions of measurement, the use of different words to describe the reflected light: reflectance factor (transmittance factor), radiance factor, reflectometer value, and so on. Because the understanding of the different terms is unimportant for color measurement, the terms will not be discussed. In the standards the reflectance factor has values between 0 and 1. In practice the term is modified by multiplying it by 100 and describing the reflectance factor in percent.

In this book the reflectance factor, used as a collective term, is indicated with the letter $R(\lambda)$ [R for reflection and (λ) as a reminder that it is wavelength-dependent].

If we measure the reflectance factor, we do not measure the energy of the incident and the energy of the reflected light. Instead, we measure the amount of light reflected from the sample and from a white standard. The ratio of the two reflected quantities is the reflectance factor. The reflectance factor therefore is measured not as an absolute value but relative to a white standard. The white standard most widely used is one that is calibrated against the perfect reflecting diffuser. A sample is called an absolute or perfect white when it diffusely reflects all of the incident light. *Diffuse reflectance* means symmetrical in all directions. It can not be produced as a real material, but in national standardization laboratories such materials can be measured in absolute terms on a theoretical basis. All real-white standards are calibrated against the theoretical absolute white. If we measure with a modern color measuring instrument, we can assume that the reflectance values are the absolute reflectance values. (If we measure transmitting samples, the standard in general is air. If we measure solutions, we measure against a cuvette filled with the solvent.)

Because the reflectance curve represents the percentage of the reflected light, the spectral power distribution of the light that illuminates the sample in the measurement is unimportant. It must contain all wavelengths of the visible spectrum in sufficient amount to generate an accurately detectable signal. This statement is not true when the samples to be measured are fluorescent, as will be discussed in detail in Chapter 4 (p. 91).

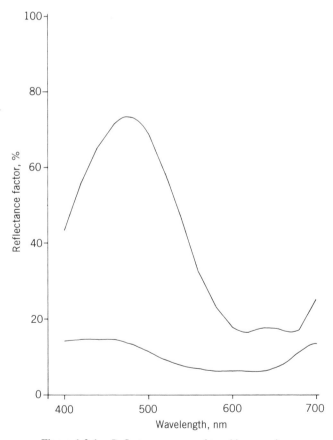

Figure 1.2-1.   Reflectance curves of two blue samples.

For the following description of the reflectance curves we assume that they are determined correctly. They are taken from the examples discussed in Chapters 2–5. The accompanying samples are textile or paper dyeings, plastic samples, or paints. The samples are colored with many different colorants as well as with different concentrations of the colorants.

Figure 1.2-1 shows the reflectance curves of two blue samples. We see that both samples reflect mostly blue light or more light of short wavelengths than yellow or red light, which is absorbed more or less strong. The samples therefore look blue, because they reflect mostly the blue light or because they absorb less blue light than light of other wavelengths. The more light they absorb or the lower the reflectance factor, the darker they look. They look more brilliant the larger the difference is between the reflectance values in the wavelength range with the highest absorption (lowest reflectance factor) and the lowest absorption (highest reflectance factor). Our curves show the reflectance curves of a dark and dull blue and that of a very brilliant blue sample.

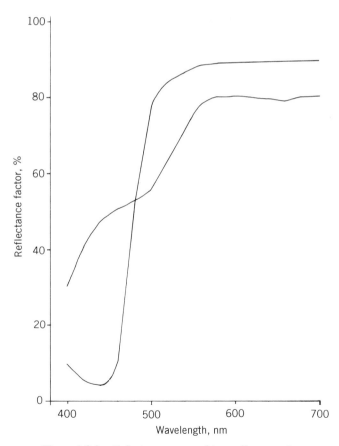

Figure 1.2-2. Reflectance curves of two yellow samples.

The curves result from the kind and the amount of colorants used, as will be discussed in detail in Chapter 5 (p. 100).

Corresponding statements are true for all other colored samples. The measured reflectance curve therefore is like a fingerprint that describes the kind and the amount of the used colorants. For the amount we can say the more colorant used the greater the absorption or the lower the reflection.

Figures 1.2-2–1.2-4 show similar curves for yellow, red, and green. The relationship between color and absorption that is described above is seen clearly. It can be seen that yellow samples reflect not only yellow light, but green, yellow, and red light to the same degree.

Figure 1.2-5 shows the reflectance curves of two theoretical ideal-gray samples. Such samples absorb all wavelengths of the incident light to the same extent. The more light absorbed, the darker the sample looks. An ideal-white sample reflects all incident light (reflectance factor 100% for all wavelengths). An ideal black sample absorbs all the incident light so that it reflects nothing (reflectance factor 0% for all wavelengths).

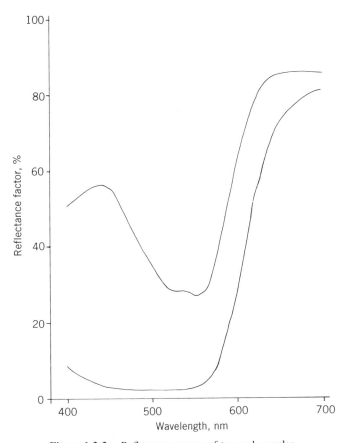

Figure 1.2-3. Reflectance curves of two red samples.

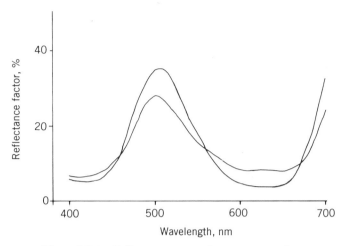

Figure 1.2-4. Reflectance curves of two green samples.

As stated above, the reflectance curve is similar to a fingerprint for the description of the color sample. Anyone who deals a little bit with color measurement will very soon be able to give a general descriptive name of the color for every reflectance curve. The name will not be very specific, but this is not necessary.

While the reflectance curves in the early days of color measurement had to be drawn by hand or, when automatically measured, at least had to be looked at for further use, with modern color measurement systems the curves are measured but often neither printed nor recorded. The user today therefore may miss a large amount of information. Because this fact is realized, the curves can be seen on the computer screen if we wish. It is hoped that the users today often use this feature.

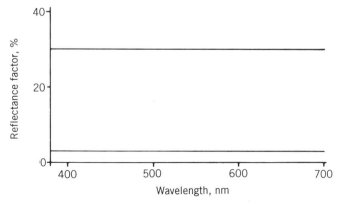

Figure 1.2-5. Reflectance curves of two ideal-gray samples.

Summary. The reflectance curve is the fingerprint of each colored sample. It can be measured objectively. It is influenced by the kind and the amount of colorants used, as will be discussed in Chapter 5 in detail. In color measurement we determine the reflectance or transmittance curve of a sample. The measured information is mathematically transformed, according to standard conventions, into the numbers we use to describe the color of the sample.

1.3. OBSERVER

The observer we now discuss is the human eye, which has receptors for the perception of color. To understand color vision, the eye has to be described briefly. To portray it in simplified terms, we can compare the eye with a camera equipped with an autofocus lens. It has, as in the camera, a variable aperture (the iris) that controls the incoming light. After the aperture is a lens with changeable focus. It displays the viewed sample focused sharply on the retina. The retina has the same function as the film in the camera. A young observer (a young eye) can view samples at a minimum distance of 3 inches, measured from the eye. The normal reading distance is 10–12 inches. (There are people for whom this statement is not true. In these individuals the lens–retina distance is different from that in a normal eye. They are short-sighted or far-sighted, and this error has to be corrected with eyeglasses.) The changing of the focus of the lens is controlled by a muscle. As people age, this muscle is not as elastic as before. Therefore the minimum distance at which we can view samples rises sharply. From the age 40–50 on it is so large that samples in the reading distance cannot be brought into focus. We become presbyopic (far-sighted) and need eyeglasses for reading. As we age more we also cannot focus on distant objects; therefore a second pair of eyeglasses (or bifocal glasses) is necessary. These interesting facts about the lens have nothing to do with color vision. But for color vision it is important that the lens, with advancing age, not only is less elastic but also develops a more or less yellow color. Therefore the light that falls on the retina is altered by the color of the lens and therefore the color perception also is changed. (In daily life this is normally not noticeable, but it is one reason why the color sensation for every human with normal color vision is different.) The effect of the aging on color vision will be discussed again in Chapter 3 (p. 62). To understand the color vision of older persons more easily, we can put a yellow filter before the eye to simulate it. The changed color vision when wearing sunglasses is generally known.

As mentioned above, the detector in the eye on which the viewed sample is projected is the retina. It contains two different kinds of light-sensitive receptors, called *cones* and *rods*. The cones are responsible for vision during the day and for color vision. The rods are responsible for vision during twilight. (There are theories of color vision stating that the rods also have a function in color vision.) Rods and cones are not uniformly distributed in the retina. There is one place in the retina that contains only cones. It is called the *fovea*. If we look at a sample, its center it always focused on the fovea. The fovea is so small that only samples are no bigger than a dime (diam. 0.7 inch) viewed at the reading distance of 10 inches are focused on the fovea. Or to state this in another way, samples with an angular subtense equal to or less than 4° are focused on the fovea. This is shown schematically in Figure 1.3-1. In the figure an angular subtense of 2° is chosen, because this viewing field is important for further discussions. A shirt button (diameter 0.4 inch) viewed at the reading distance of 10 inches corresponds to this viewing field. Outside the fovea there are cones and rods in the retina. The number of cones decreases and the number of rods increases in the outer part of the retina. The color perception outside the fovea is not the same as in the fovea. In the introduction this fact is described and can be seen infrequently without any aid—a spot of another color that moves on the sample if the head is moved. Because most of the samples are larger than 0.7 inch, the color perception in the area surrounding the fovea where parts of such samples are focused will be discussed later.

Another area of the retina should be mentioned: the "blind spot," which contains no light-sensitive receptors. The nerves that send the stimuli from the receptors to the brain are located here. For color vision it is of no significance. The blind spot is represented schematically in many textbooks.

Many theories of color vision and many papers deal with the transformation of the light falling on the retina into the stimuli that are sent to the brain. To understand color measurement, it is not absolutely necessary to study these theories.

It has been known for a long time that the color sensation can be matched by the mixing of three colored lights, or, when a sample is to be matched with colorants, generally three colorants are needed. The amount of each light or each colorant is related to the sensitivity of the receptors of the eye. We can indirectly determine the sensitivity of the receptors with such mixing experiments. These were first done in the 1920s. The trials have been done under controlled conditions. The trials further have been done independently in the laboratories of Guild and Wright. Because the trials required a lot of work, the

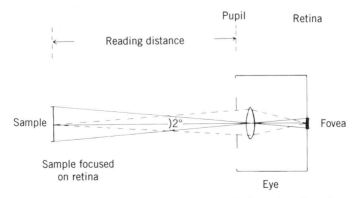

Figure 1.3-1. Schematic description of a sample focused on the retina.

number of observers was small (altogether 17). The sensitivity of the eye was measured with the help of spectral colors (these are lights whose spectral power distributions consist only of one wavelength) in principle as follows: One part of an observed screen is illuminated with light which contains only one wavelength. The wavelength of that light source is changed step by step from 380 to 780 nm. The other part is illuminated with an additive mixture of light that contains three different wavelengths: red, green, and blue (see Figure 1.3-2). The amount of each of the three spectral lights, called *primaries*, could be changed independently. For each spectral color of the visible spectrum, the amount of light of the three spectral colors that illuminate the other part of the screen can be measured. These are the amounts $\bar{r}(\lambda)$, $\bar{g}(\lambda)$, $\bar{b}(\lambda)$ of the primaries. The viewing angle was $2°$, so that light illuminates only the fovea. The result of the trials is shown in Figure 1.3-3. The negative amounts shown result from the fact that many spectral lights could be matched only when one of the primaries was added to the tested spectral light. That means that in those cases both parts of the screen are illuminated with two spectral lights.

To get such curves, in principle, many different triplets of primaries can be used. With each different triplet, we get different curves. The curves can be transformed to those obtainable with other primaries.

The curves that are measured from the different observers differ considerably, as is shown in Figures 1.3-4 and 1.3-5, which give the original measurements of the 17 observers but in a different presentation. The fact that can not be neglected is that the individual sensitivities of humans with normal color vision differ considerably. None of them has the sensitivity of the standard observer. If we describe the color of a sample or the color perception with numbers, we must always keep this in mind. How large the influence of this scattering is will be discussed in detail in Chapter 3 (p. 61).

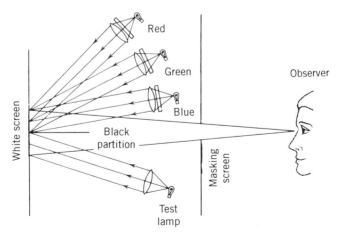

Figure 1.3-2. An arrangement for determination of the sensitivity curves (the tristimulus functions $\bar{x}$, $\bar{y}$, $\bar{z}$, or the color matching functions) of the receptors in the eye (from Billmeyer and Saltzman[1]).

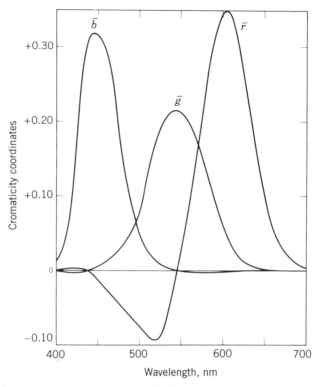

Figure 1.3-3. Color matching functions of the 2° observer (from Billmeyer and Saltzman[1]).

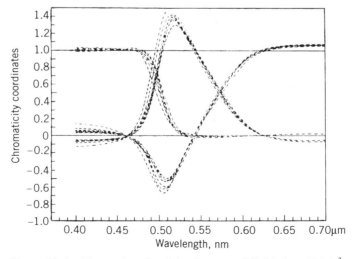

Figure 1.3-4. Measured results of the observers of Guild (from Wright[2]).

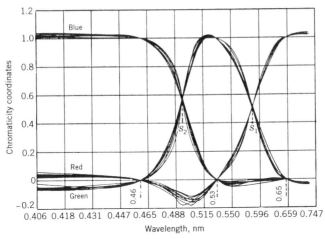

Figure 1.3-5. Measured results of the observers of Wright (from Wright[2]).

For calculating the tristimulus values of a sample, which will be described in the next section, first the color-matching functions of the standard observer shown in Figure 1.3-3 are used. They are determined from the results of the 17 observers. The observer is called the "2° observer." The CIE changed this recommendation after a few years by transforming the curves into such that would result from the use of other primaries. The chosen primaries are not real ones. They are called imaginary primaries [X], [Y], [Z] and their amounts $\bar{x}(\lambda)$, $\bar{y}(\lambda)$, $\bar{z}(\lambda)$. There were several reasons for choosing the so called imaginary primaries; for instance, all amounts of $\bar{x}(\lambda)$, $\bar{y}(\lambda)$, $\bar{z}(\lambda)$ should be positive; further, the curve $\bar{y}(\lambda)$ should be equal to the so-called spectral luminance efficiency factor of the eye, which was determined and standardized before 1931. The transformed curves of the 2° standard observer have been recommended by the CIE since 1931. They are shown in Figure 1.3-6. Using them, we can calculate what amounts of the primaries [X], [Y], [Z], are necessary to match a spectral color. In simpler speech, we call them the sensitivity curves of the three receptors in the eye. In the language of the CIE they are called CIE color-matching functions. The observer who has these color matching functions is called the CIE 1931 standard colorimetric observer or as in a shortened and understandable way the 2° standard observer.

Until here we discussed only the scattering of the sensitivities of observers with normal color vision. As stated before, not all observers have normal color vision. About 8% of the male population and only a few women are color-defective. Their color vision differs significantly from that of the remaining population. Such color vision is a defect that is transmitted as is hemophilia (recessive gene, sex-specific). People with such a defect may have one receptor that has a sensitivity significantly different from that of persons with normal vision, or one of their receptors has no sensitivity at all. Because there are three different kinds of cones in the retina, six different kinds of color defects

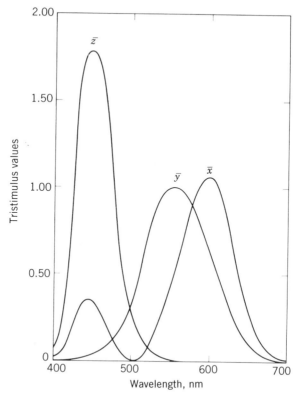

Figure 1.3-6. Color-matching functions of the 2° standard observer (from Billmeyer and Saltzman[1]).

result. People with a receptor with a different sensitivity are called *anomalous trichromats*. The *deuteranomalous* individuals have receptors for green that are different and are the largest group of color defectives. For protanomaly, the receptor for red is defective. For tritanomaly it is the receptor for blue. Tritanomaly is very seldom found. People who have only two light-sensitive receptors are called *dichromats*. There are three types: deuteranopes, protanopes, and tritanopes, representing more serious defects than the anamalous trichromats. Tritanopes are very seldom found. Deuteranopes and protanopes see only yellow and blue; they can learn to name known red and green colors correctly (McLaren,[1] Hunt[1]).

Defective color vision can be tested with color plates with "confusion" colors (pseudoisochromatic plates) or with an instrument (anomaloscope) in which the observers must make additive mixtures, as is described for the determination of the color matching functions. Unfortunately, persons who are required to make visual matches still are not generally tested for defective color vision before they start their education. This can lead to choosing the wrong type of work later.

As said before the sensitivity of the fovea is different from that of the surrounding parts of the retina. It was found, when matching pairs of larger samples, that the results of the visual matchings and the calculated color differences (Chapter 2) do not agree. Because the visual matching of larger samples in practice is more important than that of small samples, the tests that had been done to validate the 2° standard observer were repeated with a larger field of vision. An agreement was reached to make the tests with a field of view corresponding to a viewing angle of 10° (this corresponds to the viewing of a sample of about 1.9 inches at the reading distance of 10 inches). The small field corresponding to a field of view of 2° was eliminated by a mask. In this way the tests could be performed with greater ease because the fields to be matched were free of a spot with another color in them (because of the different sensitivities of the fovea and the surrounding areas). The tests were done by Stiles and Burch and by Speranskaya. Because at this time the instruments were better than in the 1920s, the number of persons tested was larger (Stiles and Burch, about 50). The scattering between the results of the different observers was not smaller than 1931. The results led to the recommendation of the 10° standard observer, which should be correctly named the "$(10 - 2)°$ observer." It was recommended in 1964 by the CIE for use in addition to the 2° standard observer. The 10° standard observer is named, in the language of the CIE, the "CIE 1964 supplementary colorimetric observer."

The curves of the 10° standard observer are drawn together with those of the 2° standard observer in Figure 1.3-7. They are called $\bar{x}_{10}(\lambda)$, $\bar{y}_{10}(\lambda)$, and $\bar{z}_{10}(\lambda)$. The differences between the two sets of curves are larger than the differences within one set of curves, which is due to the variation of the color-matching function among the individual observers.

For the calculation of the tristimulus values discussed in the next section we use one of the standard observers. In practice the use of the 10° standard observer has been adopted in most countries. The reason is that the samples in the pairs to be matched generally are larger than 0.19 inch.

In Chapter 3 the described differences in the sensitivities of the individual observers are again discussed. Because this is, as already mentioned, one limitation to the practical use of the results of color measurements. In 1989 the CIE issued the recommendation[2] for a "standard deviate observer" (ST). It is calculated from the statistical analysis of the test results of the individual observers.

Summary. The spectral sensitivity of the different receptors (cones) of the eye can be determined indirectly by trials of additive mixing.

Two sets of spectral sensitivity curves, the color matching functions of an observer who views only a small sample (light falls only on the fovea)—2° standard observer, and that of another observer who views larger samples (light falls also on the surrounding part of the retina)—10° standard observer, are recommended by the CIE.

The color matching curves of the standard observers result from the reduction of the data from the curves determined by the experiments of several workers.

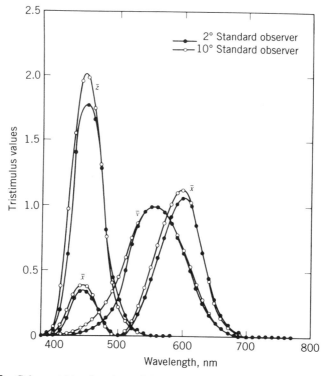

Figure 1.3-7. Color-matching functions of the 10° standard observer compared with those of the 2° standard observer (from Billmeyer and Saltzman[1]).

The number of experimental subjects was relatively small; the scattering of their results is relatively large.

The CIE recommends also a standard deviate observer to describe the influence of such scattering on the results of color measurements.

1.4. CALCULATION OF THE TRISTIMULUS VALUES

As described before, the relative spectral power distribution $S(\lambda)$ of the light that illuminates a sample can vary a great deal, and its composition is generally unknown. To get comparable values, the CIE has recommended the use of only a few standardized power distributions (standard illuminants). These are used to calculate the numbers that describe colors. The light that is reflected or transmitted by the sample is measured. This provides the values of $R(\lambda)$ or $T(\lambda)$. The sensitivity of the observer, of our eye, was previously determined indirectly. Each color sensation can be described by calculation of the amount of light that must be sent from the three imaginary primaries to the eye to get the same perception as that induced or stimulated by the light from the sample. These amounts are called tristimulus values X, Y, Z. These values describe the color perception, and the calculation of the tristimulus values is the first step for solving every task with the help of color measurement.

The measurement of the reflection curve is necessary for the calculation. The other two factors, the spectral power distribution of the standard illuminant and the color-matching functions of the standard observer, are standardized by convention. For measuring reflection curves we use spectrophotometers, which are described in detail later (p. 74). The calculation of the tristimulus values is done by computers, which are today always part of the color measurement system. The computer can do much more than only calculate the tristimulus values.

The steps for calculation are simple. They follow exactly the visual process described above. First, the light that falls on the eye is determined. It is produced by the spectral power distribution of the illuminating light source $S(\lambda)$—for calculation, that of a standard illuminant—and the part of the light that is reflected from the sample. This part is measured as the reflectance curve $R(\lambda)$. The relative power distribution that strikes the eye is the product of both values summed over the visible spectrum. Mathematically this is written

$$\sum_{\lambda} S(\lambda) \times R(\lambda) \qquad (1.4\text{-}1)$$

Lambda (λ) will be omitted under the Σ in subsequent equations.

Actually the sums are integrals, because the light that strikes the eye contains not only the wavelengths used for the sum but all wavelengths of the visible spectrum. For measurement and calculation, 20 nm intervals are normally used. The intervals can be shortened to 10 or 5 nm. The influence of the size of the intervals will be discussed later (p. 88). The summation often goes from 400 to 700 nm (or with steps of 10 or 5 nm from 380 to 700 nm).

We can imagine that the amount of light that is reflected from the sample to the eye is compared indirectly with the light that must be sent from the primaries to get the same color perception for each wavelength.

The amounts of the primaries for each wavelength are summed up to get the whole amount of light from the primaries that is needed to match the sample. We can also say that the eye determines the amount of light which is sent by the sample to it corresponding to the sensitivity of the receptors in the eye—$\bar{x}(\lambda), \bar{y}(\lambda), \bar{z}(\lambda)$ [e.g., $\bar{x}_{10}(\lambda), \bar{y}_{10}(\lambda), \bar{z}_{10}(\lambda)$]. Then it sends stimuli to the brain that are proportional to the amount of light modified by the sensitivity of the receptors. Mathematically speaking, these stimuli are identical to the amount of the primaries that are necessary to match the sample. They are called tristimulus values X, Y, Z. Shown below are the equations for calculating the tristimulus values corresponding to the description above. (They are also repeated in the Appendix.)

$$X = \sum S(\lambda) \times R(\lambda) \times \bar{x}(\lambda)$$
$$Y = \sum S(\lambda) \times R(\lambda) \times \bar{y}(\lambda) \qquad (1.4\text{-}2)$$
$$Z = \sum S(\lambda) \times R(\lambda) \times \bar{z}(\lambda)$$

In these equations $S(\lambda)$, and $\bar{x}(\lambda)$, $\bar{y}(\lambda)$, and $\bar{z}(\lambda)$ are standardized values. $R(\lambda)$ has to be measured. For calculation the products $S(\lambda) \times \bar{x}(\lambda)$, $S(\lambda) \times \bar{y}(\lambda)$, and $S(\lambda) \times \bar{z}(\lambda)$ with which we will work are stored in the computer for every illuminant, every observer, and every wavelength interval. The products are called *weighting factors*. The most extensive collection of weighting factors can be found in the American Standard ASTM E 308-85.[1]

If we go to the trouble (no user of a modern color measurement system does it!) to compare the values published at different places and the values stored in the instruments, we find small differences. The transformation of the values published by the CIE in steps of 1 nm intervals into steps of 10 or 20 nm and the inclusion of the values below 400 nm and over 700 nm is not definitely established. The calculated tristimulus values therefore can differ slightly for the same reflectance curve. For the calculation of color differences this is not significant.

The tristimulus value Y has a special significance. It is a measure of the lightness of the sample. By establishing the color matching functions of the eye (by transforming the measured values) $\bar{y}(\lambda)$ was so determined to be equal to the luminous efficiency of the eye [called $V(\lambda)$]. This is exactly true only for the 2° standard observer. The tristimulus value Y is called therefore also the luminance factor. (Although it is not absolutely correct, the tristimulus value Y for the 10° standard observer also is used as a measure of lightness.) The tristimulus values are standardized so that ideal white has the value $Y = 100$ for every illuminant and every observer (see Table 1.4-1).

If we look again at the equations for the tristimulus values

$$X = \sum S(\lambda) \times R(\lambda) \times \bar{x}(\lambda)$$

$$Y = \sum S(\lambda) \times R(\lambda) \times \bar{y}(\lambda) \qquad (1.4\text{-}2)$$

$$Z = \sum S(\lambda) \times R(\lambda) \times \bar{z}(\lambda)$$

we see clearly that the tristimulus values change for the same reflectance curve when $S(\lambda)$ or $\bar{x}(\lambda)$ . . . is changed (see also Table 1.4-1). The statement of tristimulus values therefore is definite only when the tristimulus values show clearly with what illuminant and observer they are calculated. The tristimulus values X, Y, Z always belong to the 2° standard observer; the tristimulus values X_{10}, Y_{10}, Z_{10}, always to the 10° standard observer. The illuminant mostly is given after the tristimulus value, such as $X65$ or $X_{10}A$. (Terms such as $XD65/2$ or $XA/10$ are also used; the subscript $_{10}$ is eliminated with this terminology.) Giving tristimulus values without stating the illuminant and observer is, at least, not definite, if not to say wrong. In the few cases where absolute tristimulus values are compared the exact naming of the illuminant and observer is not enough. In these special cases the weighting functions used in the calculation must be given. It is best for consistency to agree (on the weighting factors) before making the calculations.

What is stated in Section 1.3 and what we see from the equations make it

Table 1.4-1. Tristimulus Values of Samples Whose Reflectance Curves are Shown in Figures 1.2-1–1.2.5 (Tristimulus Values for Standard Illuminants D65 and A and for 2° and 10° Standard Observers Are Shown)

Sample	Illuminant D65						Illuminant A					
	X_{10}	Y_{10}	Z_{10}	X	Y	Z	X_{10}	Y_{10}	Z_{10}	X	Y	Z
Ideal white	94.81	100.00	107.30	95.04	100.00	108.88	111.14	100.00	35.20	109.85	100.00	35.59
Ideal gray 1	28.44	30.00	32.19	28.51	30.00	32.66	33.34	30.00	10.56	32.96	30.00	10.68[a]
Ideal gray 2	2.84	3.00	3.22	2.85	3.00	3.27	3.33	3.00	1.06	3.30	3.00	1.07[b]
Yellow 1	68.92	71.34	51.87	69.46	72.84	53.37	86.37	74.86	17.41	85.61	75.58	17.89
Yellow 2	70.39	81.96	18.75	70.94	84.66	21.70	94.98	86.04	8.01	94.07	87.08	9.33
Red 1	50.77	40.95	55.68	52.35	40.60	55.79	65.22	45.81	17.75	66.31	45.74	17.61
Red 2	21.13	12.02	3.73	22.79	12.62	3.64	34.70	18.23	1.15	36.45	18.58	1.13
Blue 1	30.97	41.03	71.66	30.34	38.79	73.34	27.43	33.73	23.80	26.35	32.57	24.15
Blue 2	7.68	8.21	15.03	7.65	7.85	15.14	7.74	7.45	4.86	7.63	7.29	4.85
Green 1	8.62	17.17	13.18	8.27	16.63	14.45	8.59	13.79	5.05	8.11	13.38	5.53
Green 2	10.29	15.89	12.93	10.11	15.56	13.86	11.34	13.87	4.74	10.98	13.62	5.08

[a]Values for ideal white ×0.3.
[b]Values for ideal gray ×0.1.

certain that the color can be described by three independent values; this means that color vision is three-dimensional. A color can only be plotted in a space, called the *color space*. In the color space every color has a clearly defined place.

That color perception is three-dimensional is also understandable when we try to describe a color. The statement of the hue as red or green alone does not say much about the color. It is described better when we call it a brilliant red or a dull green. Both colors can be light or dark at the same time. The statement that the color is a light brilliant blue describes the sky blue, and the statement that it is a dark dull blue describes a color often called "navy blue."

Figure 1.4-1 shows the step in the calculation of tristimulus values in flow-chart form. The single steps of calculation are clearly seen. The tristimulus values are equal to the areas under the curves of the last row.

Table 1.4-1 shows the tristimulus values of the yellow, red, green, blue, and gray samples whose reflectance curves are shown in Figures 1.2-1–1.2-5 and also that for ideal white. They are given for the standard illuminants D65 and A and as well as for the 2° and the 10° standard observers. In the moment it may be enough, to see that the three values change, when the illuminant or the observer changes.

The calculation of the tristimulus values described above are no longer done by hand today but by the computers integrated in the color measurement systems for every required illuminant–observer combination. The reflectance values necessary for the calculation are measured in the systems. Very seldom are they written during measurement as a curve. Sometimes they are shown on plotters or on the screen of the computer. They may also be printed as numbers. There are also instruments that print only the tristimulus values.

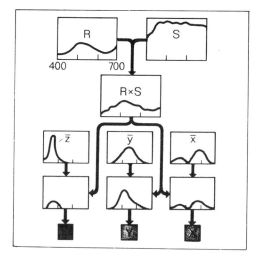

Figure 1.4-1. Interaction of the factors determining color perception (from Brockes et al.[1]).

It is possible to show the tristimulus values graphically. This can be done, as mentioned before, only in a color space. We have an agreement, to project the color space on a plane, the so-called chromaticity diagram. The coordinates of the chromaticity diagram are called *chromaticity coordinates*.

$$x = \frac{X}{X + Y + Z}$$
$$y = \frac{Y}{X + Y + Z} \qquad (1.4\text{-}3)$$

To be specific, x, y, z need the same additions (as to illuminant and observer) as the tristimulus values.

Figure 1.4-2 shows that the 1931 x, y and the 1964 x_{10}, y_{10} chromaticity diagrams are not very different. This is to be expected as they are constructed on similar principles. As the third dimension, the lightness is to be imagined perpendicular to the plane of the chromaticity diagram. The maximum possible lightness is a function of the chromaticity. Color space therefore is not a cylinder on the chromaticity diagram. Color space for all nonfluorescing colors is shown in Figures 1.4-3 and 1.4-4. (The calculation of this color space was

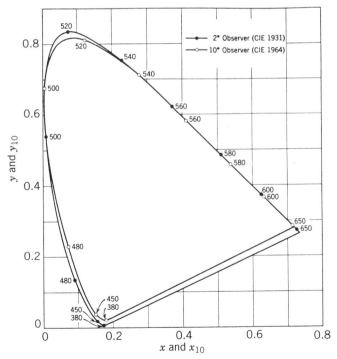

Figure 1.4-2. The 1931 x, y, and the 1964 x_{10}, y_{10} chromaticity diagrams are very similar (from Judd and Wyszecki[1]).

Figure 1.4-3. Color space in which all colors which can be produced theoretically can be found (Rösch[2]). The figure shows a model of this color space made by Rösch and that still exists.

done independently at nearly at the same time by Rösch[2] in Germany and by MacAdam[2] in the United States. The figures show two different forms of representation.) In practice we enter (if at all) only the chromaticity coordinates into the chromaticity diagram. The lightness is given as a number next to the chromaticity coordinates or in a table.

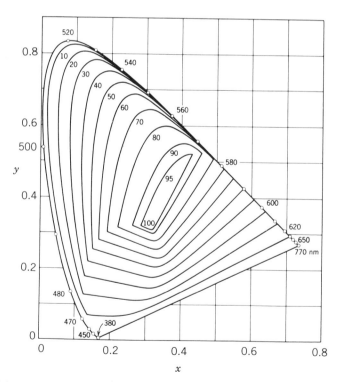

Figure 1.4-4. Color space in which all colors that theoretically can be produced are shown—drawing by MacAdam[2] (from Billmeyer and Saltzman[1]).

If we look at the chromaticity diagram more carefully, we see it has the form of a sole of a shoe. At the border of the diagram the spectral colors are to be found. All other colors are found within the sole of the shoe. Near the center of the chromaticity diagram is the point where achromatic colors are found (also called the "white point"). That means that this point has the chromaticity coordinates of all achromatic colors. If the sample is illuminated with a light source whose spectral power distribution is equal for all wavelengths, such a spectral power distribution corresponds to a color temperature of about 6000K, the chromaticity coordinates for all achromatic colors are $x = y = 0.3333$. For other illuminants we get other values, such as x D65 $= 0.3127$ and y D65 $= 0.3290$ or x_{10} D75 $= 0.2997$ and y_{10} D75 $= 0.3174$. Colors of similar hue are located on a line between the achromatic point and a spectral color. The longer the distance from the achromatic point to the chromaticity coordinates, the more brilliant (saturated) are the colors. The most saturated colors are the spectral colors.

In the chromaticity diagram described above all colors, with one exception, have one clearly defined place. The exception is black. As can be seen from equations 1.4-3 (the division by zero gives an indeterminate number), in this diagram black does not have an unambiguous place.

Table 1.4-2 shows the chromaticity coordinates x and y, as well as Y as a measure of the lightness, for the samples of Table 1.4-1 for standard illuminant D65 and the 10° standard observer.

The chromaticity diagram is not the only way of representing chromaticity graphically. In the discussion of the calculation of color differences (p. 35) we will discuss other methods. We work much more today with plots other than the CIE chromaticity diagram.

Table 1.4-2. Chromaticity Coordinates and Luminance Factor for the Samples from Table 1.4-1 for Standard Illuminant D65 and 10° Standard Observer

	x_{10}/D65	y_{10}/D65	Y_{10}/D65
Ideal white	0.3138	0.3310	100.00
Ideal gray 1	0.3138	0.3310	30.00
Ideal gray 2	0.3138	0.3310	3.00
Yellow 1	0.3587	0.3713	71.34
Yellow 2	0.4114	0.4791	81.96
Red 1	0.3444	0.2778	40.95
Red 2	0.5730	0.3259	12.02
Blue 1	0.2156	0.2655	41.03
Blue 2	0.2483	0.2655	8.21
Green 1	0.2212	0.4407	17.17
Green 2	0.2631	0.4063	15.89

Summary

1. The color of a sample is described by the tristimulus values X, Y, Z.
2. The tristimulus values are composed of the products of the factors described below summed for all wavelengths:

   The spectral power distribution of the light that illuminates the sample $S(\lambda)$. The calculations are always done with a standard illuminant (e.g., D65) or with an illuminant recommended for use by the CIE (e.g., F11). (*Warning*: For visual matching the light illuminating the samples never corresponds with a standard illuminant.)

   The spectral reflectance factor of the sample to be matched. This is measured with a spectrophotometer. (*Warning*: The precision of modern color measurement instruments is very high. The measured sample (or more exactly the measured part of the sample) is, for many reasons, which will be discussed later (p. 146), often not representative. For this reason the result may be subject to error.)

   The spectral sensitivity of the receptors in the eye, which is standardized: 2° standard observer, 10° standard observer. (*Warning*: There is no observer who has the spectral sensitivities that correspond to that of a standard observer. Additionally the size of the samples to be matched generally does not correspond with that which was used for the determination of the standard observer.)

3. Independent of all warnings, the following statement is correct. Two samples with the same reflectance curve $R_1 = R_2$ look the same to all observers and for all illuminants or light sources. They also have the same tristimulus values.

$$X = \sum S_L(\lambda) \times R_1(\lambda) \times \bar{x}_B(\lambda) = \sum S_L(\lambda) \times R_2(\lambda) \times \bar{x}_B(\lambda)$$

$$Y = \sum S_L(\lambda) \times R_1(\lambda) \times \bar{y}_B(\lambda) = \sum S_L(\lambda) \times R_2(\lambda) \times \bar{y}_B(\lambda) \quad (1.4\text{-}4)$$

$$Z = \sum S_L(\lambda) \times R_1(\lambda) \times \bar{z}_B(\lambda) = \sum S_L(\lambda) \times R_2(\lambda) \times \bar{z}_B(\lambda)$$

For $R_1 = R_2$ the sums are equal regardless of which illuminant (S_L) or what observer ($\bar{x}_B$, $\bar{y}_B$, $\bar{z}_B$) is used in the calculation.

Although the concept of metamerism should be described first, because it shows what happens when the reflectance curves of two samples are different, I will first discuss the determination of color differences (Chapter 2, p. 35). With the terms learned in Chapter 2, metamerism is easier to describe and to understand.

2

Calculation of Color Difference

As described in detail in Chapter 1, the sensation of color can be described with the help of numbers called tristimulus values. If these values are the same for a pair of samples at one matching condition (illuminant–observer) and the pair of samples is nonmetameric (see next chapter), both samples look the same to any observer and under any light source. In practice very few samples absolutely fulfill this condition. Also samples that should theoretically be the same may have substantial differences in their tristimulus values and also in the perceived color sensation. (This is the case, for example, if we measure different parts of a sample that may not be uniform in color or if we produce a sample under the "same" conditions a second time.)

In order to describe color differences with numbers, a very large amount of experimental data has been accumulated. Such numbers have a very large practical importance for quality control, arrangements with buyers on the color difference tolerance between standard and delivery, calculation of the metamerism index (see p. 70), and so on.

The problem could be easily solved if the same distances in the color space, where the coordinates are the tristimulus values, would correspond to the same visual color differences. Unfortunately that is not true.

The search for a color difference equation therefore started immediately after the CIE system was recommended in 1931. All trials in principle have the same objective. The color space in which the coordinates length, width, and height are the tristimulus values X, Y, Z is transformed (distorted) in such a way that the color space that results from this procedure shows the same linear distances for the same visual color differences. The distance between two samples in a visual uniform color space corresponding to the color difference between the two samples is always called ΔE (Δ is the mathematical symbol for difference). In the calculation of color difference, ΔE, with the help of all color difference formulas, the numbers of the standard always are subtracted from the numbers of the sample (sample minus standard).

Because, as stated before, many transformations of color space exist, the statement of a value ΔE is meaningless without describing how it is obtained or calculated. (This is an error that is unfortunately made very often, and that will be pointed out again and again.)

In mathematics the distance between two points in a space with the coordinates l,m,n is described by the following equation:

$$\Delta E = [(\Delta l)^2 + (\Delta m)^2 + (\Delta n)^2]^{1/2} \tag{2-1}$$

If the color space is visually uniform, ΔE is the color difference sought. All color difference formulas have the general form described above. However, they describe the differences in different ways in the various color spaces.

If we calculate color differences it is useful to calculate not only the overall color difference but also to look for the individual components $\Delta l, \Delta m, \Delta n$. It need not be $\Delta l, \Delta m, \Delta n$; it could also be values derived from them. These numbers give valuable information as to the nature of the differences and to what might be done to correct them. The same overall color difference may some time be within the tolerance limit and another time outside the limit.

The problem of the transformation of the CIE color space to a visually uniform space has been worked on in many different ways.

The experiments of MacAdam in the early 1940s are justly famous. He tested the threshold sensitivity of the eye to chromaticity differences in the matching of colored lights. He found that the threshold could be described with ellipses on the chromaticity diagram. The work was later extended (with Brown) to combined lightness and chromaticity and the results were expressed in three-dimensional ellipsoids.

Figure 2-1 shows the famous MacAdam ellipses corresponding to 10 times the standard deviation of color matching. The ellipses show the matching results of one observer. Tests that were done later with more observers show that the ellipses for every observer differ in shape and size.

The same statement is true, when in contrast to the very small color differences, examined by MacAdam, larger color differences are judged by different observers. This is one essential reason why the problem of the assessment of color differences is not yet solved to the satisfaction of all. It will probably never be solved to everyone's satisfaction.

The results of the trials of MacAdam were later described mathematically so the calculation of color differences was possible. First it was done mostly graphically and later, together with new results, in a form slightly changed with the help of equations. The FMC-2 (Friele, MacAdam, Chickering) color difference formula is the only formula used, up to now, which is based on the MacAdam ellipses. When calculations are done with a color difference formula based on the MacAdam ellipses, the color difference ΔE is given in MacAdam units. Without saying which formula is used and without saying how the calculations are done, the meaning of the MacAdam unit is not clear.

The FMC-2 formula is given in detail in the Appendix (p. 164). Because the ellipsoids are transformed into spheres of the same size, this formula is complicated. The calculation of color differences with the help of this formula therefore can be done only with the help of a computer; with a computer it is not difficult. (This applies also to all of the other formulas currently in use.)

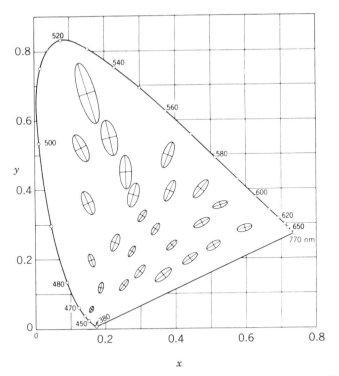

Figure 2-1. Chromaticity diagram for the 2° standard observer. The MacAdam ellipses corresponding to 10 times the standard deviation of color matching for a single observer are shown. (From Billmeyer and Saltzman.[1])

Although the FMC-2 formula is used less today, it is compared with more modern color difference formulas in a few examples. Many of the older and important papers that describe applications give data in MacAdam units. Many modern textbooks make reference to the formula but do not show it. I am of the opinion that it does not hurt to cite a historically important formula and also to discuss it.

FMC-2 formula:

$$\Delta E = [(\Delta L)^2 + (\Delta C_{r\text{-}g})^2 + \Delta C_{y\text{-}b})^2]^{1/2} \tag{2-2}$$

where L = lightness
        $C_{r\text{-}g}, C_{y\text{-}b}$ = chroma in the corresponding color space

Transformations of the CIE system were also made with the help of different theories of color vision. One of the resulting color spaces was published by the NBS (U.S. National Bureau of Standards—now the National Institute of Standards and Technology, NIST). The color difference calculated in this space therefore is given in NBS units. Before 1970 many color difference formulas were "converted" to so-called NBS units by the use of a conversion factor. Without an exact statement as to which formula is used for the calculation of the given color differences, it is uncertain what the so-called NBS units mean.

Further transformations had the goal to transform the CIE systems so that the color chips of the Munsell atlas give equal color differences between adjacent chips. (The Munsell color atlas was first published in 1929). It shows a collection of color ships with visually uniform spacing between chips. The Munsell color atlas is the most widely used color atlas today, especially in the United States. In contrast to the experiments on the threshold sensitivity of the eye, which show a sensitivity to a color difference that is smaller than a practical tolerance, the samples in the Munsell atlas have color differences between one sample and the other that are larger than most commercial tolerances.

One formula developed from such examination and used to a large extent in earlier days is the Adams–Nickerson formula. In a modified form it is called ANLAB 40 formula. Again, as CIELAB, with a further small modification it was recommended in 1976 by the CIE.

Up to the early 1970s a large number of color difference formulas were published. A few of them were widely used, as in connection with filter colorimeters. Some three filter colorimeters (see Chapter 4) gave, as the result of the measurement, not the CIE tristimulus values but the coordinates of a transformed color space. The color difference could then be calculated (with the help of formula 2-1) with the simple calculators that were available at that time. Because the use of direct-reading colorimeters reduced the calculations required, this was, in the time before computers, the only practical possibility to do objective quality control.

The published color difference equations were, as previously stated, developed partly from different theories; in part, they were only modifications of known formulas. In the latter case the goal was to allow simpler calculations while giving better results. Because no equation was standardized, all users of color measurements used the formula they liked best for different reasons.

The determined color difference was given either in MacAdam units or in NBS units (or without any unit). In most cases the color difference formula used for the calculations was not given. Today this is also often the case. In the early 1970s there was a great deal of confusion concerning color differences. Talks between colleagues, talks between seller and buyer, and numbers in publications seemed to come from a modern version of the Tower of Babel. No one understood the others.

It was shown at this time also, as is confirmed below with examples, that the color difference calculated with one formula could not be transformed to the color difference calculated with another formula by a factor. If one decides to calculate color difference with several formulas, the color differences have to be calculated separately for each formula starting with the measured tristimulus values. (For the modern computers such calculations are not a problem.)

Despite this knowledge the CIE decided very late to recommend one (or several) formulas for general use. All known formulas had some disadvantages balanced by some advantages. The scattering of the visual judgments was very large; none of the formulas gave equal color differences for pairs of samples that are visually judged to have the same color difference on the average. To put it in another way this means that at the time of the recommendations no agreement had been reached on a visually uniform color space. This statement is still true today, although in the last few years considerable progress has been made. There are many reasons to doubt that the solution of the problem in a general valid form ever will be found or agreed on.

Although none of the formulas known met the requirement for uniformity, in response to worldwide demand a common language had to be found and the CIE decided in 1976 to recommend two color spaces for practical use. Starting from these color spaces the CIE recommended two color difference formulas. The formulas are *CIELUV* and *CIELAB*. In practice in most cases the CIELAB formula is used. The CIELUV formula is used only in special cases, when additive mixtures are to be judged. (The CIELUV formula is described in most of the textbooks in the Bibliography. It is also stored in some of the color measurement systems. Readers interested in the use of this formula are advised to evaluate their samples with this formula.)

The *L*, a*, b** (CIELAB) color space and the color differences that result from this color space are described with the following equations. (Coordinates from standardized color spaces and the resulting color differences are marked with an asterisk.)

CIELAB formula:

$$\Delta E^* = [(\Delta a^*)^2 + (\Delta b^*)^2 + (\Delta L^*)^2]^{1/2} \qquad (2\text{-}3)$$
$$\Delta E^* = [(\Delta L^*)^2 + (\Delta C^*)^2 + (\Delta H^*)^2]^{1/2}$$

where L^* = Lightness
   a^*, b^* = chroma coordinates
       C^* = chroma
    ΔH^* = hue difference

The complete set of equations is presented in the Appendix (p. 165).

The transformation of the tristimulus values X, Y, Z into the values L^*, a^*, b^* is relatively simple. The factor of standardization is chosen so that a color difference that is satisfactory gives a value of ~ 1. This statement is only a clue because the color difference that is satisfactory is, to a significant degree, dependent on the specific product. Further, the CIELAB color space is not visually uniform. Empirical knowledge tells us that, for example, special saturated yellow colors may have a larger value for the tolerance than unsaturated colors of low chroma.

Figure 2-2 shows the L^*, a^*, b^* color space. The derived quantities saturation (chroma) C^* and hue h are also indicated. L^* may have values between 0 and 100. a^* and b^* may have values between around -80 and $+80$. Colors with no chroma always have the value $a^* = b^* = 0$. Because the opponent color theory is used to develop the transformation, one of the coordinates (a^*) shows the redness or the greenness of the color the other coordinate (b^*) shows the yellowness or the blueness. (This theory states that the color sensation originates by addition or subtraction of the stimuli which are sent to the brain (Chapter 1). The greenness and the blueness are given with negative sign. In this color space the lightness is perpendicular to the plane of the a^*, b^* diagram. The a^*, b^* diagram differs from the CIE chromaticity diagram in an important quality. The colors with no chroma are to be found, for every illuminant, at the zero point of the plane (intersection of the two axes). This is achieved by dividing the tristimulus values by the tristimulus values of the ideal white (for the same illumination-observer combination). This is shown in the equations in the Appendix (p. 165).

In Figures 2-3 and 2-4 the resolution of the overall color difference (ΔE) into Δa^* and Δb^* and in the color differences ΔC^* (chroma difference) and ΔH^* (hue difference), respectively, is shown. There are computer programs that show this information on the screen.

The search for a uniform color space was and is not finished with the recommendation of CIELAB and CIELUV. Because of the wide distribution of color measurement systems and that of fast computers with larger capacity, more measurements of visually evaluated samples are available, and we have larger calculation programs to analyze the data. Today several more modern

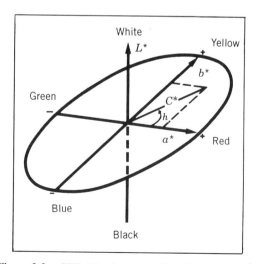

Figure 2-2. CIELAB color space (from Brockes et al.[1])

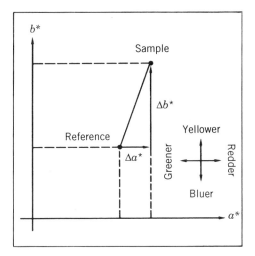

Figure 2-3. Resolution of the chromaticity difference into Δa^* and Δb^* in the CIELAB system (from Brockes et al.[1]).

formulas are used (Up to now none of them is recommended by the CIE to replace CIELAB, but it may be expected in the near future.)

The uniformity of the newer formulas is significantly better. All formulas modify the values ΔL^*, ΔC^*, and ΔH^* calculated with the CIELAB equations with the help of factors. (This is also true for the unpublished formula from Marks & Spencer, as can be shown by indirect analyses.)

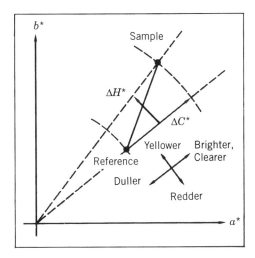

Figure 2-4. Resolution of the chromaticity difference into ΔC^* and ΔH^* in the CIELAB system (from Brockes et al.[1]).

The best known and most widely distributed of them is the CMC $(l:c)$ formula (Colour Measurement Committee, Society of Dyers and Colourists, Great Britain) which is given below. (This is the most recent version of the formula, as published by McDonald.[1]) The complete formula is given in the Appendix. It is expected that this formula soon will be recommended by the CIE.)

CMC $(l:c)$ formula:

$$\Delta E = \left[\left(\frac{\Delta L^*}{lS_L}\right)^2 + \left(\frac{\Delta C^*}{cS_C}\right)^2 + \left(\frac{\Delta H^*}{S_H}\right)^2\right]^{1/2} \qquad (2\text{-}4)$$

where S_L = function of L
S_C = function of C
S_H = function of H and C
l, c = correction factors which can be chosen so that the calculated numbers correspond to optimal values for certain matching conditions or certain samples.

The analysis of all the published matching data shows that the combination $(2:1)$ where 2 is shorthand for $2(S_L)$ and 1 is the factor for $1(S_C)$, gives the best correlation with the visual matching results for acceptable color tolerances (pass/fail decisions, for quality control). If one analyzes the trials in which the threshold sensitivity of the eye is compared with color differences, the combination $(1:1)$ gives the better correlation. With the latter combination the difference in the lightness has a higher influence than with the combination $(2:1)$.

In Figure 2-5 the correlations between the published matching data and the CIELAB or the CMC $(l:c)$ values are shown graphically. It can be seen that the correlation with the CMC $(l:c)$ formula is significantly better. The resulting scattering is still large because of the scattering of the results from visual matchings. The reasons for this are the differences between the eyes of observers; this factor is especially important for matching metameric samples. Personal preferences are also a factor in matching color differences. Also, measuring errors should not be neglected.

Independent of the color difference formula used, it is not enough to calculate the color difference for only one illuminant–observer condition and to assume that this difference is the same for every other illuminant and for every other observer. (As described in Chapter 1 [p. 21] the actual observer is never the standard observer.) The calculation has to be done each time for the conditions that correspond best with the real matching conditions.

In spite of all the progress in the determination of color differences, all the positive experience in quality control and statistical quality protection, and the value of prior agreement between buyer and seller, all the previous warnings are now summarized:

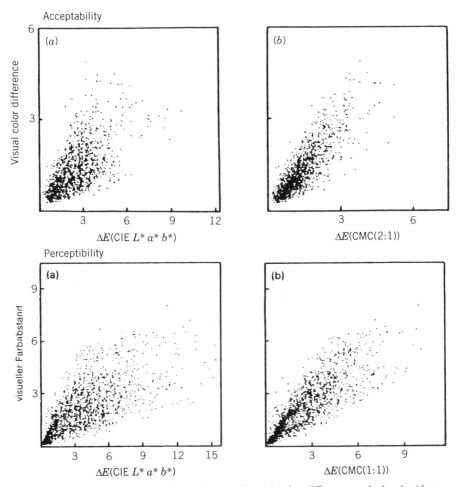

Figure 2-5. Correlation of visual matching results and color differences calculated with two formulas; all published data are used (from Luo and Rigg[2]).

1. The statement of a color difference is meaningless without specifying the formula that is used for the calculation.
2. Color differences that are calculated with one formula cannot be transformed into those calculated with another formula by simply using a factor.
3. The color difference tolerance is dependent on the product being tested (e.g., for automotive paints and knitting yarns the color differences must be very small).
4. There is no color difference formula that is absolutely uniform.
5. The light source for matching is always different from the standard illuminant used for the calculation, and the observer always has a sensi-

tivity that does not correspond with the sensitivity of the standard observer.

This is one reason why the matching results of individual observers differ from each other. They do not correspond in size or in the nature of the difference. For this reason calculations must be done for the conditions that agree as closely as possible with the conditions of the visual matchings.

This is particularly true when the color difference for the pairs of samples to be matched are at the border of the tolerance limit and when the samples are metameric; this is true also when the matches are being judged under the same light source.

In summary, in spite of the unquestionable advantages of measurement data, the result should be tested visually. When it is borderline, more than one judgment should be made.

Examples at the end of this chapter are given which reinforce what has been said above. Mostly only the overall color difference is discussed.

To avoid confusion the CMC formula is used only in the form CMC (2:1). As stated above (p. 42), the difference in lightness in this form adds less to the overall color difference than in the case of CMC (1:1). The difference between the results with the CMC (2:1) and (1:1) versions is small. The results with both formulas differ considerably from that with the CIELAB formula because of the correction for chroma. As far as it is possible to test formulas during practical use, I have found over the years that the CMC (2:1) formula fulfills the demand for equal tolerances in all of color space much better than the CIELAB formula.

Table 2-1 shows the color differences for 6 pairs of samples, which are tested for quality control (paper dyeings). While the dyeings have only small differences in color strength (see Chapter 5) the small differences in shade are more important. The color of the standards are shown in Figure 2-6. For clearer illustration they are drawn in the a^*, b^*, chroma diagram. The L^* values are given beneath the number of the pair of samples. The calculations are done with standard illuminant D65 and the $10°$ standard observer. In addition, the calculations are done with all three equations discussed above: FMC-2, CIE-

Table 2-1. Color Differences for Samples with a Small Color Difference. Calculated with D65/10.

Pair of Samples	CIELAB	ΔE CMC (2:1)	FMC-2
1	0.23	0.11	0.60
2	1.48	0.50	2.58
3	0.85	0.39	2.69
4	0.80	0.46	2.03
5	3.06	2.06	9.00
6	0.24	0.26	0.92

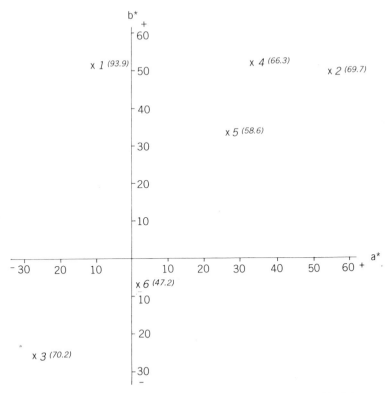

Figure 2-6. a^*, b^*, L^* (in brackets) for the samples from Table 2-1.

LAB, and CMC (2-1). Because the color differences in these examples are very small, they are given to two decimal places, in order to check the factors (Figure 2-7). Otherwise the statement of the second decimal is meaningless, as is the color difference because the samples never can be measured so accurately; also the accuracy of the visual matching is much smaller. Because a picture is clearer than a table, the factors between the formulas that are calculated from the calculated color differences are shown in a Figure 2-7.

Apart from showing that color differences calculated with the FMC-2 formula in general are much larger and that the color differences calculated with the CMC (2:1) formula are in general smaller than those calculated with the CIELAB formula, the warning not to work with factors is clearly demonstrated.

Table 2-2 shows the color differences of pairs of samples from a commission dyehouse (for textiles). The colors of the standards are shown in Figure 2-8, which otherwise is equivalent to Figure 2-6. The color differences are larger and the pairs of samples have a considerable amount of metamerism. They also show large differences in color strength. This example will be discussed again below as well as in Chapter 3, "Metamerism" (p. 70) and again in the discussion of the precision of the results of color measurements (preparation of

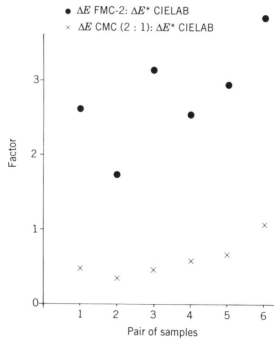

Figure 2-7. Ratio of the color differences FMC-2/CIELAB and CMC (2:1)/CIELAB from Table 2-1.

Table 2-2. Color Differences for Metameric Textile Samples with Commercial Color Differences. Calculated for D65/10.

Pair of Samples	CIELAB	ΔE CMC (2:1)	FMC-2
1	2.5	1.6	8.8
2	2.0	0.9	6.4
3	1.4	0.9	4.2
4	1.5	0.7	3.7
5	3.0	1.3	6.0
6	1.9	1.5	4.3
7	1.7	1.0	7.0
8	0.9	0.7	4.6
9	1.0	1.4	3.2
10	2.6	1.5	9.7
11	5.8	2.4	15.0
12	2.8	1.5	9.8
13	1.9	1.3	4.6
14	2.1	1.1	4.9
15	0.6	0.3	1.4

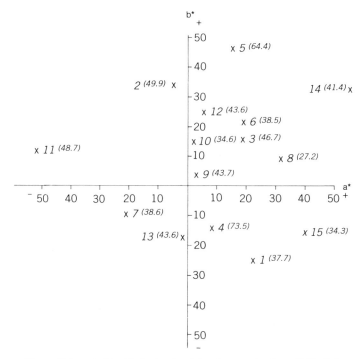

Figure 2-8. $a*$, $b*$, $L*$ (in brackets) for the samples from Table 2-2.

the sample and precision of color measurement) (p. 157). The calculations are again made for D65 and the 10° standard observer. The factors between the formulas, shown in Figure 2-9, scatter as much as in the first example (Figures 2-6, 2-7, Table 2-1).

Color difference measurements are designed, in principle, for nonmetameric pairs of samples (see Chapter 3), as should always be true for quality control. In computer color matching (see Chapter 5 and the examples in this chapter) metamerism cannot always be avoided. In quality control the match accepted by the customer, never the original submission, should then become the standard. This will help avoid metamerism in production and measurement.

In Table 2-3 the overall total color differences are shown again, but this time also split up in the three components that show the differences in lightness, chroma, and hue. Although this table may be confusing because of the large amount of numbers, it is worth careful examination. Of interest are not the absolute values, but the ratio of the three components ΔL, ΔC, and ΔH or ΔL, Δa and Δb, and the influence of each of the three variables on the overall color difference. We can see that the CMC (2 : 1) formula, ΔL and ΔC have a smaller influence on the overall color difference than for the CIELAB formula. Therefore, the influence of the hue difference is larger for the CMC (2 : 1) formula. If we look at CIELAB and FMC-2 we can see that also the kind of calculated

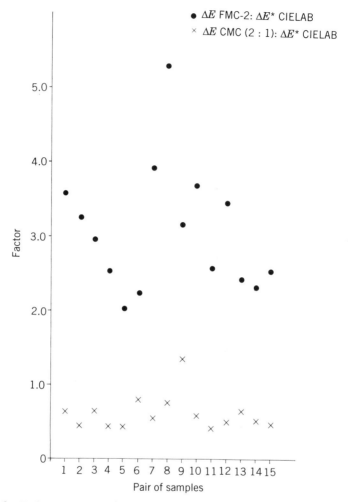

Figure 2-9. Ratio of the color differences FMC-2/CIELAB and CMC (2:1)/CIELAB from Table 2-2.

differences are not equal (sign and ratio of the values Δa^* and Δb^* or $\Delta C_{r\text{-}g}$ and $\Delta C_{y\text{-}b}$).

In the discussion of the CMC $(l:c)$ formula it was emphasized that the correlation with visual matchings also for this formula is not perfect. Therefore, again there is a warning not to blindly believe the measured and calculated numbers, but to check them visually, especially when the measured samples are to be corrected, as is done for production control or process control. If the correction is not calculated (see Chapter 5), it is recommended, if possible, before correcting, to look at the reflectance curves of reference and sample.

Table 2-4 and Figures 2-10, and 2-11 show the color differences between paint samples, which are prepared for visual tests of color differences. The

Table 2-3. Splitting Up of the Color Difference for the Samples from Table 2-2

Pair of Samples	Formula	a^*	b^*	L^*	ΔE^* or ΔE	Δa^* / ΔC_{r-g}	Δb^* / ΔC_{y-b}	ΔL^* / ΔL_{corr}	ΔC^* / ΔC_{corr}	ΔH^* / ΔH_{corr}
1	CIELAB	22.0	-25.6	37.6	2.5	-2.0	-0.4	1.4	-1.0	-1.8
	CMC(2:1)				1.6			0.8	-0.5	-1.3
	FMC-2				8.8			3.4		
2	CIELAB	-6.1	33.1	49.9	2.0	-8.2	0.4	0.6	-1.8	-0.2
	CMC(2:1)				0.9	0.5	-1.8	0.3	-0.3	-0.1
	FMC-2				6.4			1.6		
3	CIELAB	16.3	16.7	46.7	1.4	-0.1	-6.2	-1.2	0.4	-0.5
	CMC(2:1)				0.9	0.6	-0.1	-0.6	0.2	0.7
	FMC-2				4.2			-2.9		
4	CIELAB	7.7	-13.5	73.5	1.5	2.9	1.0	1.4	0.3	0.3
	CMC(2:1)				0.7	0.4	-0.1	0.5	0.2	0.3
	FMC-2				3.7			3.6		
5	CIELAB	15.2	45.6	64.4	3.0	1.0	0.1	-1.4	-2.6	0.7
	CMC(2:1)				1.3	-1.5	-2.3	-0.5	-1.0	0.6
	FMC-2				6.0			-3.1		
6	CIELAB	18.0	-22.4	38.5	1.9	-4.3	-2.8	1.2	1.1	-1.0
	CMC(2:1)				1.5	1.5	0.3	0.6	0.6	-1.3
	FMC-2				4.3			2.9		
7	CIELAB	-22.3	-9.3	38.6	1.7	3.1	-0.7	-0.2	1.5	0.8
	CMC(2:1)				1.0	-1.2	-1.3	-0.1	0.8	0.5
	FMC-2				7.0			-0.3		
8	CIELAB	30.9	8.2	27.2	0.9	-4.5	-5.3	0.4	0.4	0.8
	CMC(2:1)				0.7	0.1	0.9	0.3	0.2	0.6
	FMC-2				4.6			0.9		
9	CIELAB	2.7	4.3	43.7	0.9	0.5	4.5	0.3	-0.3	-0.8
	CMC(2:1)				1.4	0.6	-0.8	0.2	-0.3	-1.3
	FMC-2				3.2	0.3	-3.1	0.8		

Table 2-3. (*Continued*)

Pair of Samples	Formula	a^*	b^*	L^* or	ΔE^* / ΔE	Δa^* / ΔC_{r-g}	Δb^* / ΔC_{y-b}	ΔL^* / ΔL_{corr}	ΔC^* / ΔC_{corr}	ΔH^* / ΔH_{corr}
10	CIELAB	1.9	15.0	34.6	2.6	-0.4	-0.3	2.6	-0.3	0.3
	CMC (2:1)				1.5			1.5	-0.2	0.4
	FMC-2				9.7	-2.7	-6.7	6.6		
11	CIELAB	-52.1	11.3	48.7	5.8	4.6	-1.7	3.1	-4.8	0.8
	CMC (2:1)				2.4			1.4	-1.8	0.4
	FMC-2				15.0	10.7	-7.8	6.9		
12	CIELAB	5.5	24.6	43.6	2.8	0.3	-0.5	2.7	-0.4	-0.4
	CMC (2:1)				1.5			1.4	-0.2	-0.4
	FMC-2				9.8	-1.4	-6.7	7.1		
13	CIELAB	-0.9	-17.3	33.8	1.9	0.7	-0.5	1.7	0.5	0.7
	CMC (2:1)				1.3			1.0	0.3	0.7
	FMC-2				4.6	2.0	0.3	4.1		
14	CIELAB	56.2	32.9	41.4	2.1	-1.0	0.5	-1.8	-0.6	0.9
	CMC (2:1)				1.1			-0.9	-0.2	0.6
	FMC-2				4.9	2.2	3.1	-3.1		
15	CIELAB	40.2	-16.7	34.3	0.6	0.3	-0.4	0.3	0.4	-0.2
	CMC (2:1)				0.3			0.2	0.2	-0.1
	FMC-2				1.4	-0.4	-1.2	0.7		

Table 2-4. Color Differences for Paints[a]

Pair of samples	CIELAB	ΔE CMC (2:1)	FMC-2
1	2.6	1.1	6.5
2	1.2	1.0	2.1
3	1.8	0.9	4.4
4	0.7	0.4	1.4
5	1.1	0.7	2.5
6	1.8	1.1	3.5
7	2.1	1.2	6.4
8	2.3	1.3	3.3
9	1.7	0.9	5.4
10	2.2	1.1	5.6
11	1.0	0.5	3.3
12	2.0	1.0	4.8

[a]Measurement with gloss included. Gloss corrected by calculation. Differences shown here are mostly differences in color strength. Calculated for D65/10.

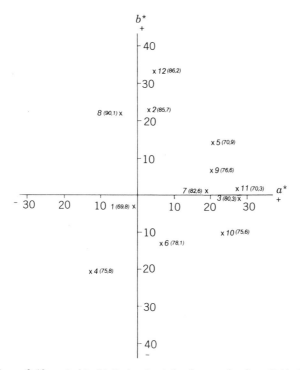

Figure 2-10. a^*, b^*, L^* (in brackets) for the samples from Table 2-4.

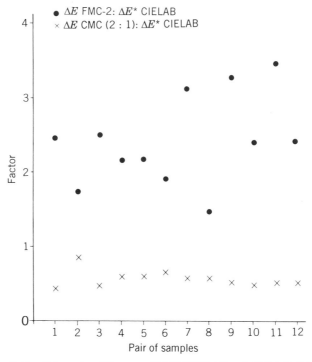

Figure 2-11. Ratio of the color differences FMC-2/CIELAB and CMC (2 : 1)/CIELAB from Table 2-4.

table and the figures correspond to those of the two examples discussed above. The pairs of samples differed mostly in the concentration of the colorants. They are measured with gloss included (see Chapter 4). Because glossy samples always are matched for color without gloss as much as possible, the gloss was corrected by calculation. The results correspond to the results of the two other examples, as they should, if the color difference formulas discussed above are generally valid.

The statement that color differences that are calculated with one illuminant–observer condition cannot be transferred to another illuminant–observer condition is confirmed with one example.

Table 2-5 shows the color differences for the textile samples and the paint samples that are, on average, equally large, calculated also for other illuminant–observer conditions. Because I got for a few pairs of samples only the tristimulus values X, Y, Z for D65/10, I could not calculate the color differences for other conditions. This is the reason why they are missing in the table.

The calculations for the textile samples also have to be done with different measurement results for the same samples. The attentive reader will see that the color differences for D65/10 differ in an extreme case by more than $\Delta E = 1$, a fact discussed in detail in Chapter 6. The calculations have been done

Table 2-5. Color Difference of the Same Pairs of Samples Discussed in Tables 2-2 and 2-4 with Different Formulas[a]

Pair of Samples	ΔE CMC (2:1)						
	D65/10	D75/10	A/10	F2/10	F11/10	D65/2	D65/10ST
Paint							
1	1.03	1.02	1.07	1.06	1.05	1.03	1.02
2	1.00	1.02	0.68	0.69	0.80	0.92	0.95
3	0.83	0.84	0.76	0.71	0.70	0.79	0.80
4	0.38	0.39	0.35	0.35	0.46	0.50	0.40
5	0.66	0.67	0.68	0.87	0.71	0.71	0.67
6	1.14	1.14	1.17	1.20	1.21	1.15	1.13
7	1.16	1.16	1.14	1.00	1.16	1.17	1.15
8	1.26	1.26	1.28	1.32	1.33	1.30	1.26
9	0.80	0.79	0.85	0.71	0.83	0.84	0.81
11	0.45	0.44	0.51	0.40	0.45	0.45	0.45
Textile							
1	3.46	3.47	3.38	3.50	3.71	3.44	3.44
2	0.68	1.14	4.26	0.83	3.00	1.60	0.36
8	0.72	0.71	1.03	1.56	1.05	0.86	0.73
9	1.46	1.55	0.82	0.91	1.42	1.12	1.32
10	1.93	2.04	2.81	1.58	1.59	1.88	1.88
11	2.68	2.84	1.88	3.10	2.59	2.20	2.42
13	1.22	1.26	1.12	1.18	1.15	1.15	1.19
14	0.99	1.02	0.93	1.03	1.09	0.94	0.95
15	0.59	0.56	0.74	0.63	0.62	0.72	0.62

[a]This time the calculations are done only with the CMC (2:1) formula, but with different illuminant–observer combinations.

with the CMC $(2:1)$ formula. Both the illuminant and the observer were changed. In the table the color differences that differ by more than 20% from those calculated with D65/10 are underlined. In extreme cases the differences can be 500%. If we look at the individual illumination–observer conditions in this table, we see the following:

D75/10 and D65/10 differ by only 1000K. Daylight can change within a few minutes by this amount. Besides that there are many light booths which have a color temperature of 7500K. Already with such a small change the color difference varies, for strongly metameric samples, by more than 20% (see Chapter 3).

A/10 (incandescent light) differs much more in color temperature: therefore, the color differences often change by more than 20%

F2/10 (fluorescent lamp) and F11/10 (prime color fluorescent lamp), both with a relative spectral power distribution that differs strongly from that of D65 and also with a correlated color temperature of about 4000K, also show, as expected, large changes in several cases.

In addition to the relative spectral power distribution of the light source, the nature of the observer also has an influence on the color difference.

Between D65/2 and D65/10 a few pairs of samples show differences larger than 20% between the calculated color differences. That means that small samples should be judged with the 2° observer and larger samples with the 10° observer. Most of the samples to be matched will be larger than the size that corresponds to the 2° observer (both samples together are approximately 0.7 inch); therefore, calculations with the 10° observer generally will give better results. The samples to be matched in general will have the size that corresponds to neither the 2° observer nor the 10° observer.

The scattering of the results of every single observer can be described with the CIE standard deviate observer (ST). The standard deviate observer describes the average difference of the color difference for a group of observers to that of the standard observer. The difference between two observers may be twice as much (or larger; see Chapter 3, p. 167). Therefore all differences between D65/10 and D65/ST that are larger than 10% are underlined.

In starting with measurements and calculations for quality control, the color differences should first be calculated with several formulas and when the matching conditions are not clear, also with several illuminant–observer conditions. Today this is easily done because most of the color measurement systems can calculate the required results.

The results of the calculations should be compared with the results of visual matching. When there is enough experience, the formula that on average gives the best agreement should be used. The size of the tolerance limits is also set in this way. To a high degree this value depends on the product to be tested and sometimes on the color of the product and the nature of the color difference (whether it is in hue, value, or saturation). Several sellers of color measurement systems provide the possibility of adapting the CIELAB formula with the help of factors to one's own matching results and to set one's own limits for the color tolerance. The next step then is a pass/fail control of the production. The results of such a procedure are clearly shown on the computer screen (Figure 2-12). If there is not enough experience (comparison of the measured and calculated results with the results of visual matching), such methods should not be used, because the result that is so clearly shown may not agree with visual judgment.

For comparing color differences that are determined at different places (tolerance in contracts of supply), the color difference formula has to be agreed on, so as to speak the same language at the different places.

Summary. On the basis of theoretical considerations and/or experimental trials, scientists have been successful in transforming the color space given by the tristimulus values X, Y, Z into other color spaces that are more visually uniform. In the transformed color spaces the distance between two samples corresponds to the visual color difference. When the tristimulus values of the two samples are known, the transformed values give each of the samples a definite place in the transformed color space.

Until today no transformation equation (and no color difference formula, which is derived from the transformed color space) has been found that com-

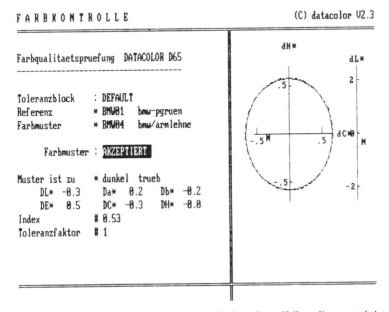

Figure 2-12. Information on the computer screen display of pass/fail quality-control data. The calculations are done for the 10° standard observer.

pletely fulfills the requirement for uniformity. The reasons are found either in the scattering of the visual matching results or in the measuring errors of the matched samples. Further, the real methods of matching do not agree with the conditions of the calculation.

The CIE recommends at the moment the CIELAB and the CIELUV formula. We can expect a change in the recommendations in the near future. Probably the CIELAB formula will be replaced by the CMC $(l:c)$ formula.

In general, the calculations are done with the illuminant–observer condition D65/10.

The color difference values that are calculated with one formula cannot be transformed, by the use of factors, to the values calculated with another formula.

With reservation it can be said that color differences $\Delta E = 1$ in CIELAB units and one of $\Delta E = 0.5$ in CMC $(2:1)$ units can be seen clearly. For some products these values are also the magnitude of the tolerance limit.

Although the number of color difference formulas (used in measurement systems) has been much smaller since 1976, a given color difference is meaningless without the statement of the formula used for the calculations.

3

Metamerism

We discussed the calculation of the tristimulus values in Chapter 1. It was shown that two samples with the same reflectance curve match under all conditions of illumination and observation. They also have the same tristimulus values. The opposite case, where two samples with different reflectance curves look different under all conditions, is not valid. There are samples with different reflectance curves that match at least under one illumination and for one observer. Such pairs of samples are called *metameric matches, metameric pairs*, or simply *metamers*. This chapter deals in some detail with the phenomenon of metamerism.

In Chapter 1 it was explained that the tristimulus values that describe the color sensation are sums (or, more exactly, integrals). We sum up the products of $S(\lambda) \times R(\lambda) \times \bar{x}(\lambda)$, $S(\lambda) \times R(\lambda)\bar{y}(\lambda)$, $S(\lambda) \times R(\lambda) \times \bar{z}(\lambda)$ (normally from 16 values at 16 wavelengths). Just as similar sums can be calculated from different numbers

$$10 = 1 + 2 + 7 = 2 + 3 + 5 = 4 + 5 + 1 = 1 + 6 + 3 \ldots$$

in the same way the same tristimulus values can be calculated from different reflectance curves, provided they fulfill certain conditions.

In Figure 3-1 curves are shown that fulfill such conditions. The reflectance curves of two gray samples are displayed. Sample 1 is an ideal gray. Sample 2 is a gray that is metameric to sample 1.

In Figure 3-2 the weighting factors $S(\lambda) \times \bar{x}(\lambda)$, $S(\lambda) \times \bar{y}(\lambda)$, and $S(\lambda) \times \bar{z}(\lambda)$ are shown for illuminant D65 and the 10° standard observer. The values of reflectance must be multiplied by these numbers, to get the tristimulus values. By comparing the reflectance curves it is seen that sample 2 shows reflectance factors that are larger or smaller at wavelengths that are shorter than the maxima of the curves of the weighting factors; at wavelengths that are longer than the maxima, the ratios are opposite. The calculated sums are equal for all three tristimulus values for both samples. The samples are metameric for D65 and the 10° standard observer. They match under these conditions.

Figure 3-3 shows the curves of the weighting factors for standard illuminant A and the 10° standard observer. It can be seen that these curves clearly differ from the curves in Figure 3-2—the maxima of the curves are clearly shifted to longer wavelengths. With these curves that is, under this illuminant–observer

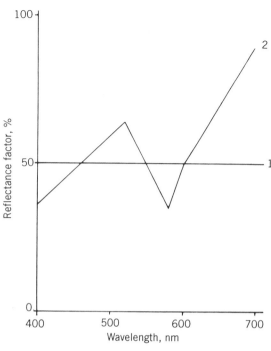

Figure 3-1. Pair of gray samples, which are metameric for D65 and the 10° standard observer.

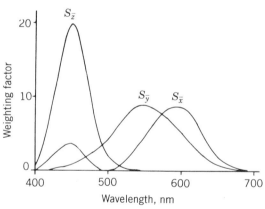

Figure 3-2. Weighting factors for D65/10.

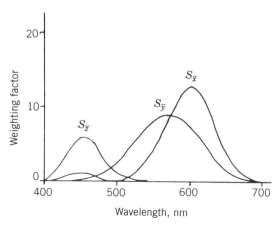

Figure 3-3. Weighting factors for A/10.

condition, different tristimulus values are calculated for samples 1 and 2. The pair of samples shows a color difference under these conditions as is seen in Table 3-1 where the tristimulus values and the color differences of the two samples under both conditions are given.

The curves of samples 1 and 2 cross at three wavelengths. There are many papers that show that a metameric pair of samples must at least cross at three wavelengths. This is, in general, true, and the intersections are a characteristic mark of metameric pairs of samples. Because the exception proves the rule, there are also papers describing metameric pairs of samples that cross at only two points.

Because matching of metameric pairs of samples is one of the most important reasons for discrepancies between the results of color measurement and the visual sensation, metamerism is discussed below in great detail with the help of many examples.

The color rule of Davidson & Hemmendinger is particularly suitable for demonstrating metamerism. (D&H color rule, now called the MatchPoint™ Metameric Color Rule, is available from Macbeth Div., Kollmorgan Corp., Newburgh, NY, 12550.) The new version, however, is made with different colorants and may give results that are not the same as those obtained with the original. Dr. Hemmendinger kindly furnished the reflectance curves of Figures 3-4 and 3-5 as well as the measured reflectance values of the original D&H color rule.

The D&H color rule consists of two strips of painted samples. The reflectance curves of the samples are shown in Figure 3-4. One strip is marked with letters and the other strip is marked with numbers.

The samples on each strip are arranged side by side. The lettered strip is above the numbered strip as in a slide rule. Both strips can be moved separately against each other. One sample of each strip can be seen in a mask. To match the colors of the two scales, the two strips are moved against each other until the two visible samples match. They should show no color difference or one that is as small as possible.

When several observers make a match under the same light source, the results often differ considerably. Similar results are obtained when one observer makes matches under several light sources. The repeatability of the test is

Table 3-1. Tristimulus Values and Color Differences of the Pair of Metameric Grays Under Two Conditions of Illumination Which are Illustrated in Figures 3-1 to 3-3

		X	Y	Z	ΔE^* CIELAB
Ideal gray (1)	D65/10	47.41	50.00	53.65	0.2
Metameric gray (2)		47.40	50.07	53.68	
Ideal gray (1)	A/10	55.57	50.00	17.60	6.5
Metameric gray (2)		58.35	50.04	17.97	

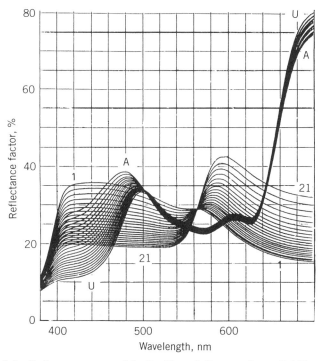

Figure 3-4. Reflectance curves of the Davidson & Hemmendinger (D&H) color rule.

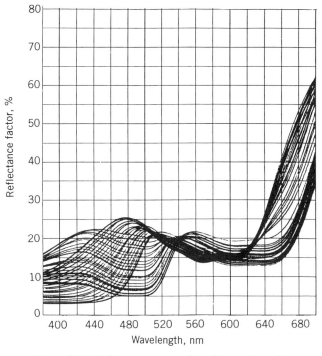

Figure 3-5. Reflectance curves of the Glenn color rule.

excellent. When one observer does the same test several times, the deviation of the results is not larger than half a step.

A second, similar color rule was made with textile samples (Glenn color rule). It is unfortunately not available today. Because results that were obtained with this rule are also discussed, Figure 3-5 shows the reflectance curves of this rule.

Figures 3-6 through 3-8 show the match results obtained with both rules.

Figure 3-6 (Berger[2]) shows results obtained with the Glenn color rule; 36 observers with normal color vision did the test under two light sources. One light source was a high-pressure xenon arc lamp (simulated daylight); the second was incandescent light. The large scatter of the match results between the observers under one light source is evident. The shifting of the match points, when the light source is changed, is also clearly seen. From the results one can calculate how large the color difference is between two samples for one observer, which match for a second observer (see below). What is obtained with the color rules, that is, with samples with large metamerism, is encountered every day when matching pairs of samples in industry, because the samples to be matched are metameric to a greater or lesser degree.

Figure 3-7 shows the relationship between the age of the observer and the results of the matching. The figure shows that such a relationship in fact exists, but the scatter is large. One reason for the influence of age is the yellowing of the lens of the eye, which is described in Chapter 1. Because of the yellowing of the lens, the light that strikes the retina of an older observer is yellower than that striking the retina of a younger observer. If the match is done in daylight, the light that goes to the retina of an older observer is similar to the light of an incandescent lamp (the artificial illumination in the evening). Because of

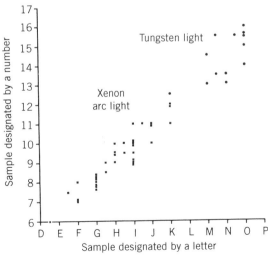

Figure 3-6. Matching results with the Glenn color rule (from Berger[2]).

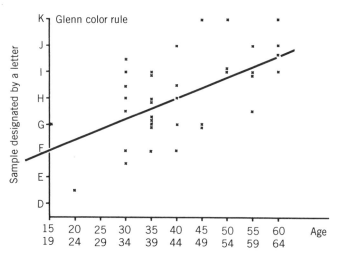

Figure 3-7. Matching results with the Glenn color rule as a function of age (from Berger[2]).

the type of scatter of the data it is unlikely that the yellowing of the lens of the eye can alone be responsible for the differences in the sensitivity of the retina of different observers. Other factors such as changes in the crystallinity of the lens may also contribute to such variation.

Figure 3-8 (Billmeyer and Saltzman[1]) shows the results of similar matchings with the D&H color rule. The range of age of the observers is similar to the range of age of the observers from Berger. The matchings were done under artificial daylight (Macbeth, 6500K) and Horizon Daylight, which is yellower than standard illuminant A. The scattering of the results is of the same magnitude as that shown in Figure 3-6. The test results shown in Figure 3-8 were repeated by Nardi with students who ranged in age between 17 and 29 years. The scattering is reduced to a quarter of that described by Billmeyer and Saltzman.

The results shown in Figures 3-9 to 3-11 were calculated with the help of the reflectance curves of Figure 3-4. The calculations were done for different observers and different illuminants (D65/10, D65/2, A/10). The calculated results are drawn in a sector of the $a*$, $b*$ diagram. The lightness is not given because the differences are small and therefore of secondary significance. It is clear that the crossing point of the two curves are at a different combination of letter and number for each illuminant–observer conditions shown. For D65/10 the pair of samples F/7 is considered a match. The test results of Billmeyer and Saltzman show as a mean of their observers the value G/8. The agreement is remarkably good, because their observers were not standard observers and their illumination was only similar to standard illuminant D65. The size of the samples does not correspond to either the 2° or the 10° standard observer.

One can further calculate how large the color difference of the pair of samples F/7 is for another observer–illuminant condition. The color difference ΔE

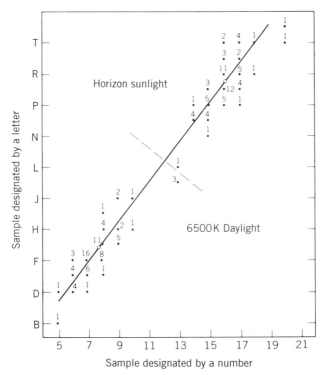

Figure 3-8. Matching results with the Davidson & Hemmendinger color rule (from Billmeyer and Saltzman[1]). The number next to the points in the figure indicates how many observers called this point their "match point."

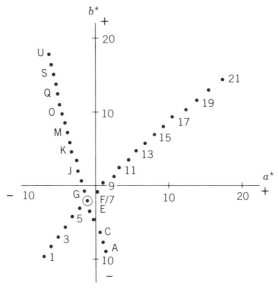

Figure 3-9. Calculated a^*, b^* values for the samples of the Davidson & Hemmendinger color rule. Standard illuminant D65, 10° standard observer.

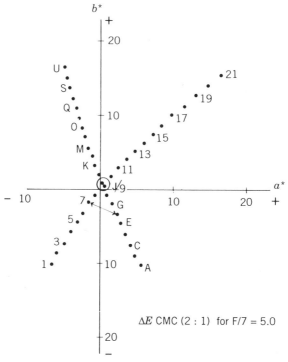

Figure 3-10. Calculated a^*, b^* values for the samples of the Davidson & Hemmendinger color rule. Standard illuminant D65, 2° standard observer.

CMC (2 : 1) is given in Figures 3-10 and 3-11. The color difference between two samples within a strip is around 1.9. One can see that the color difference due to the change in illuminant as well as the color difference due to observer variation is much greater than a normal delivery tolerance.

Figure 3-12 shows the match points for additional illuminants and the 10° standard observer. It is not surprising that the match points for illuminants D75, C, XE2 (unfiltered xenon light), and XE5 (filtered xenon light)—XE2 and XE5—see Wyszecki[2]) are not far from the match points calculated for D65. What is surprising and unexpected is that the three fluorescent lamps F2 (4230K), F7 (6500K), and F11 (4000K), which are recommended for use by the CIE, intersect near D65. All three have very different spectral power distributions and correlated color temperatures that differ also from D65. This is in contradiction to Table 2-5 and also to other results discussed below. It means that in the examination of artificial matching lamps for agreement with daylight, it is not sufficient to use only a color rule. It must be done in more than one way. The CIE has recommended the use of five pairs of metameric samples (see below and Wyszecki and Stiles[1]) to examine the spectral power distribution of artificial daylight lamps. A further measure for conformance is the color-rendering index, which is discussed in Chapter 1.

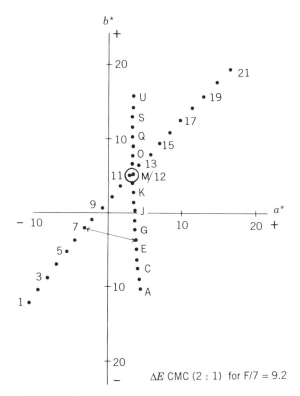

Figure 3-11. Calculated a^*, b^* values for the samples of the Davidson & Hemmendinger color rule. Standard illuminant A, 10° standard observer.

Figure 3-13 shows the match points for the 10° and the 2° standard observers, as well as for the 10° standard deviate observers 10° ST, 10° STA (old), and 10° STJ (young). The deviate observers STA and STJ were determined just as the standard deviate observer described above (p. 54). From the universe of observers only those within a specific age range were chosen for the calculations. It is important to note again the large difference between the 2° and 10° standard observers and the fact that the size of the samples to be matched, in general, is not in accordance with either of these conditions.

It should also be noted that the difference between the match points 10° STA and 10° STJ is, as expected, twice as large as the difference between 10° and 10° ST. The group of 20 observers selected by the CIE for the calculation of the standard deviate observer despite all their efforts may not be representative. The differences found by Billmeyer and Saltzman are about 7.5 time as large as those calculated with the help of the standard deviate observer. This means that calculations of color differences that are done with the help of the standard deviate observer should be multiplied by 4 ($\pm 2\sigma$) to get the difference between the calculated color difference and that visible to two real observers.

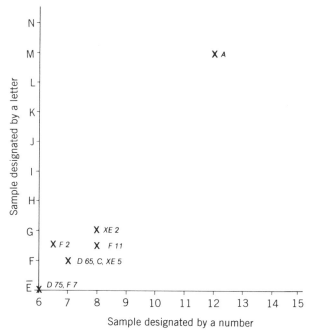

Figure 3-12. Matching pairs of samples (results rounded each time to 0.5 step) of the Davidson & Hemmendinger color rule for the 10° standard observer and different illuminants.

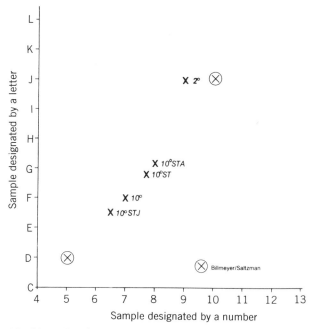

Figure 3-13. Matching pairs of samples of the Davidson & Hemmendinger color rule for standard illuminant D65 and different observers. This figure also shows the limits of match points of the observers from Figure 3-8 (for Macbeth Daylight D 6500). The difference between these limits is about 7.5 times as large as the difference between the 10° standard observer (10°) and the 10° standard deviate observer (10° ST).

The scattering found by Berger with the Glenn color rule (Figure 3-6) are of the same magnitude.

Because metamerism is so important further examples are discussed. Figure 3-14 shows the reflectance curves of the five pairs of samples recommended by the CIE to test lamps for agreement with D65 in the visible spectrum. It must be admitted that the metamerism of the pairs of samples is also larger than usually found in commercial matches. These samples were chosen from dyed textile samples that were dyed to test the agreement of the metamerism index with the result of matchings. (Today these old samples no longer exist; the agreement of artificial daylight with real daylight can no longer be done with the help of these samples. It is possible only by calculation with the recommended reflectance factors.) The results in Table 3-2 correspond with those that, in unfavorable cases, can be found in daily life. In Table 3-2 the metamerism indices ΔE^* CIELAB between these pairs of samples are given for different illuminant–observer conditions.

The color difference calculated for a selected illuminant–observer condition, which is different from the condition for which the pair of samples match, is called the *metamerism index* (MI). The statement of a metamerism index needs the statement of the illuminant–observer condition for which the pair of samples match and also the statement of the illuminant–observer condition for which the metamerism index is calculated.

In Table 3-2 the influence of the light source used for matching and of the observer is clearly seen. The average of the five values is given; that average is recommended by the CIE as a measure of the quality of the agreement of an artificial light source with D65. Lamps with values <0.5 are called "very good" (highest quality) and those with values >2.0 are called "bad."

Table 3-2 also shows that the three artificial light sources with a color temperature about 6500K (F7, XE2, XE5) group within the two highest-quality classes. That is not surprising for the xenon arc lamps. As was stated in Chapter 1, F7 is a fluorescent lamp classified as "good" using other criteria such as the color-rendering index and the color rule. The light sources with different color temperatures (A, F2, F11—about 3000–4000K) have, as expected, larger metamerism indices. F2 and F11 are grouped differently with this test as compared with results using the D&H color rule. The two observers D65/2 and D65/10ST give large metamerism indices. If they were lamps, they would be classified as "bad."

In the last example of this group it is shown that the calculated color difference (metamerism index) is self-evident from the examination of the differences in the reflectance curves. Table 3-3 shows the metamerism indices for seven samples, which are equal to the standard pattern for D65/10. They are taken from the same series of dyeings, which were chosen by the CIE (see Figure 3-14) for the testing of daylight lamps. It is evident that the color differences of the pairs of samples for the different illuminant–observer conditions differ in size.

The ratio of the color differences also changes. Specifically, a pair of samples

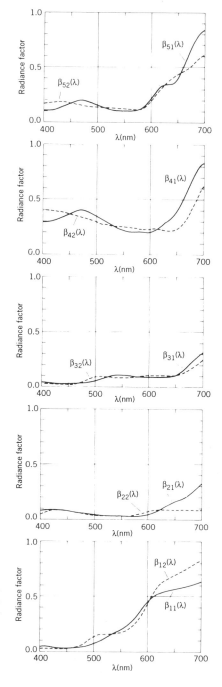

Figure 3-14. Reflectance curves of the five pairs of samples to be used, as recommended by the CIE, for the testing of artificial daylight lamps for agreement with D65. (From Wyszecki and Stiles.[1]) The reflectance curves are, in accordance with CIE, marked with β and not, as in this book, simplified with R.

Table 3-2. Metamerism Indices for the Five Pairs of Samples from Figure 3-14 for Different Illuminant–Observer Conditions[a]

Illuminant–Observer	ΔE^* CIELAB					
	1	2	3	4	5	Average
D75/10	0.5	0.3	0.5	0.7	0.4	0.5
C/10	0.3	0.3	0.2	0.2	0.4	0.3
F7/10	0.4	0.6	0.7	0.7	0.9	0.6
XE2/10	0.9	1.0	0.3	0.9	0.1	0.6
XE5/10	0.4	0.4	0.2	0.3	0.3	0.3
A/10	4.2	3.6	2.7	5.6	2.0	3.6
F2/10	2.2	2.8	2.6	1.5	1.8	2.2
F11/10	2.9	3.0	6.0	1.5	3.7	3.4
D65/2	2.6	1.4	2.3	3.3	2.1	2.3
D65/10 ST	0.7	0.5	0.4	0.9	0.6	0.6

[a]For D65/10 the pairs of samples show no color difference.

that has a smaller metamerism index than does another pair for one illuminant–observer condition can have a larger metamerism index than that same pair for another illuminant–observer condition. It means also that one cannot know how large a calculated metamerism index is for other than the calculated conditions. It has to be calculated for the specific illuminant–observer condition of interest.

Table 3-3. Metamerism Indices for Seven Pairs of Samples with the Same Standard for Different Illuminant–Observer Conditions[a]

Illuminant–Observer	ΔE CMC (2:1)						
	1	2	3	4	5	6	7
D75/10	0.4	0.5	0.4	0.1	0.2	0.7	0.2
C/10	0.1	0.2	0.1	0.0	0.2	0.1	0.2
F7/10	0.7	0.2	0.2	0.3	0.5	0.4	0.4
XE2/10	0.4	1.0	0.6	0.4	0.1	1.0	0.1
XE5/10	0.2	0.2	0.2	0.1	0.2	0.4	0.1
A/10	3.1	5.2	3.5	1.8	1.3	6.2	1.4
F2/10	2.7	1.4	1.1	0.9	1.1	1.6	1.2
F11/10	6.5	1.3	3.2	1.0	1.6	4.0	2.2
D65/2	1.7	1.8	1.5	0.7	0.9	2.6	0.9
D65/10 ST	0.3	0.5	0.4	0.1	0.3	0.8	0.2

[a]For D65/10 the pairs of samples show no color difference.

In the examples described above the color difference of the pair of samples under test was zero under the illuminant–observer conditions that were chosen as standard. The color differences under other illuminant–observer conditions is due only to metamerism. The color difference for the "nonstandard" illuminant–observer condition is called the *metamerism index* (MI).

If a standard is matched with the help of a color formulation system, the conditions are similar. The computer calculates (where possible) a formula of colorants that match the standard at the chosen illuminant–observer condition. This means that the formula gives almost the same tristimulus values and the match with the calculated formulas has no color difference from the standard. Together with the formula, the metamerism indices calculated for other illuminant–observer conditions are printed out. If any printed formulas show a color difference, the corrected metamerism index (see below) is printed. When recipes are calculated with the help of a color formulation system it is almost always possible to calculate recipes that show only a small metamerism index. To get those formulas, however, sometimes colorants that are not among those normally used, or those that are expensive, must be chosen. This is one reason why, in practice, matches with considerable metamerism are still made, as shown by the textile samples in Table 3-4.

When measurements are made on real samples, such as between the standard and the proposed match or standard and production sample (between both almost always) a small color difference will almost always be found at the chosen condition of illuminant and observer. If, in this case, one calculates the color difference for another illuminant–observer condition, the color difference at the standard condition goes into the result as an error. The CIE recommends the removal of this error by the calculation of metamerism indices with the help of a correction factor. Often the multiplicative correction is used. It must be emphasized, however, that both the size and nature of the color difference at the standard condition have a strong influence on the usefulness of a metam-

Table 3-4. Metamerism Indices of the Textile Dyeings Discussed in Chapter 2 (See Table 2-5)[a]

Illuminant–Observer	ΔE CMC $(2:1)$								
	1	2	8	9	10	11	13	14	15
D75/10	0.1	0.5	0.0	0.2	0.3	0.2	0.1	0.1	0.0
A/10	0.5	5.0	0.4	1.3	2.5	1.9	0.3	0.5	0.2
F2/10	0.5	1.1	0.7	0.6	0.4	0.6	0.3	0.0	0.3
F11/10	0.9	3.5	0.4	0.9	0.5	1.0	0.5	0.5	0.2
D65/2	0.2	2.2	0.2	0.4	1.0	0.8	0.2	0.2	0.2
D65/10 ST	0.1	0.6	0.1	0.2	0.5	0.4	0.1	0.1	0.0

[a]The metamerism index is applied to D65/10.

erism index calculated with the help of a multiplicative or other correction. In extreme cases the error can be as large as 30%. The calculations for the multiplicative correction and an additive correction are given in the Appendix (pp. 166–168).

Table 3-4 shows the corrected metamerism indices calculated with the help of the multiplicative correction [CMC (2 : 1)] of the nine pairs of textile samples from Table 2-5. These values are (as expected) generally smaller than the values in the examples described before. On the other hand, a few of them are so large that the visual matching of these pairs of samples will cause difficulties. There may be a dispute between buyer and seller as to whether the preestablished tolerances are met.

It is difficult or impossible to understand why—despite the ability, with the aid of computer color-matching programs to minimize or even eliminate metamerism—highly metameric matches are made. Any possible savings due to lower colorant costs, in my opinion, is completely offset by the difficulties caused by such matches. Problems are encountered early, during the dyeing trials in the laboratory. They require more trials and more time, therefore more expense. Further there will always be disagreements; seller and buyer will have differing opinions as to how well the match is made.

Summary. A pair of samples with identical reflectance curves appear the same under all light sources and for all observers (nonmetameric samples).

A pair of samples with different reflectance curves may look the same for one source and for one observer. Such a pair of samples is called *metameric*.

A metameric pair of samples usually match under only one condition of illuminant–observer. Under other illuminant–observer conditions a color difference is seen. The amount of the color difference depends on the size of the difference between the two reflectance curves. (There are exceptions, which are of mostly theoretical interest. It is possible to calculate metameric reflectance curves and even produce some samples that match under two or three illuminant–observer conditions.)

The color difference of a metameric pair of samples at other illumination–observer conditions is called the *metamerism index*. The statement of a metamerism index requires the conditions under which the pair of samples match and the statement of the illuminant–observer condition for which it is calculated.

To calculate a metamerism index a correction is sometimes necessary to eliminate the color difference of the pair of samples that exist under the standard condition. The quality of the correction depends on the amount of metamerism and the size of the color difference between the pair of samples under the standard conditions. The error of the metamerism index calculated with the help of a correction factor can be as high as 30% in extreme cases.

With the help of a computer color matching program it is possible to choose colorants to match standards that show little metamerism. In practice this possibility should be used more often, because the reduction of the costs of colorants will be offset through other costs and trouble. The buyer today has the chance to order matches with only a small metamerism index.

For quality control, the first production lot or the laboratory sample agreed on by the customer should always be used as standard. In that case the estimation of the color difference is determined with a nonmetameric pair of samples.

4

Color Measurement Systems; Measurement of Fluorescent Samples and Whiteness

We discussed in Chapter 1 that color measurement, in principle, is nothing else than the measuring of the reflectance curve, transmittance curve, or both. The tristimulus values and the values derived from them as color difference and metamerism index are calculated with the help of the measured reflectance factors. If we measure the reflectance curve of a pattern, it is also possible to calculate recipes. By coloring the material with the calculated colorant concentrations the pattern can be matched. This is discussed in Chapter 5 (p. 127). Apart from the standardized assumptions about the illuminant, color matching functions of the observer, a uniform color space, the relationship between reflectance, and the concentration of the colorants, the result of color measurement is dependent only on the accuracy of the measured reflectance curve.

For color measurement we need a color measuring instrument, generally a spectrophotometer, and a computer. Color measurement systems consist of these two components.

The computer can be part of the spectrophotometer or freestanding. The computer is used in the calibration of the spectrophotometer. In addition it can, at least, calculate the tristimulus values and the metamerism index for more than one illuminant–observer condition. (Macbeth has in its instruments besides the standardized illuminant–observer combinations also the weighting factors for its light booths. Therefore the tristimulus values can be calculated for the real matching conditions, but only for the standard observer.) Further they are able to calculate color differences often with more than one color difference formula. In many cases the computer is large enough that computer color matching can be done. Often further tasks are performed, such as the printing of dyeing forms in connection with the stock-taking of colorants, chemicals, and materials needed for the coloring of the product. The computer, which is updated almost every year with a faster and more efficient model here is taken for granted. The software is of primary importance for the most productive use of the computerized system. For the development of efficient software wide knowledge of colorimetry as well as of the different production processes is necessary. For this reason there are only a small number of suppliers of color measuring systems (see list in Section 3 of Bibliography) in the world. Often they adapt standard software to the special needs of the user.

Color measurement systems from new suppliers must be thoroughly tested. There are cases known where the software has so many errors that not even the tristimulus values or the color differences are calculated correctly. The large producers of colorants are a reliable source of information about the color measurement systems on the market.

We will discuss (in detail) the spectrophotometer, which has an important influence on the accuracy of color measurement.

4.1. SPECTROPHOTOMETERS

When we speak in this section of spectrophotometers we must specify that most of the spectrophotometers used today are not continuous-measuring spectrophotometers. They are the so-called abridged spectrophotometers. They are able to measure only the reflectance factors for certain wavelengths (usually 16 or 31) that are set by the instrument and not continuously for all wavelengths of the visible spectrum. Continuous-recording spectrophotometers for color measurement are sold by only a few companies. They are used mostly in research laboratories. Most of the abridged spectrophotometers used today are not as big as the older ones. But they are still generally so large and so heavy that a place for measurements has to be installed. The samples have to be taken to this place for measurement by the instrument. This means that the size of a sample is limited. To provide the ability for measuring larger samples, producers of spectrophotometers build instruments where the measuring head is a separate part connected with fiber optics and electrical wires with the main instrument. In such cases the measuring head can be put on the sample to be measured. Recently transportable instruments have been introduced; they can be brought to the samples. The possibilities for color measurement are therefore increased (art objects, historical monuments, cars). In certain spectrophotometer models, instruments are connected with a portable computer, and both the instrument and the computer can be brought to the sample. Recently a small spectrophotometer ($3^3/_{16} \times 3 \times 7^3/_4$ inches) has become available that can measure and store 500 reflectance curves.

Figure 4.1-1 is a schematic diagram of the light path in the instrument. In contrast with the eye, which judges the incoming light for all wavelengths at the same time, the measurement of the reflectance curve has to be done wavelength by wavelength. Therefore it is necessary to divide the light from the light source into the separate wavelengths. This can be done before the light falls on the sample, monochromatic illumination; or the reflected light can be separated after hitting the sample, polychromatic illumination. Most modern instruments have polychromatic illumination, because only in this way can fluorescent samples be measured accurately. This is discussed in detail in Section 4.4.

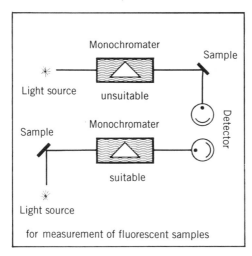

Figure 4.1-1. Scheme of the light path in a color measuring instrument (from Brockes et al.[1])

4.1.1. Light Sources

As stated earlier in discussion of the reflectance curve, the light used for the measurement must contain all wavelengths of the visible spectrum. The spectral power distribution of this source is not important unless fluorescent samples are to be measured. If fluorescent samples are to be measured so that their reflectance curves and the calculated tristimulus values agree with the results of visual matchings, the light source used for illumination of the sample must have a spectral power distribution similar to that of the light used when looking at the match.

The evaluation of the spectral power distribution of a light source in the color measuring instrument is difficult. It can be tested by the user, if at all, only indirectly. As already stated in Chapter 1, we need special instruments for that. The measurement of the spectral power distribution of the light source itself is not enough. The radiation of the light source is changed in the instrument by absorption before striking the sample. When we measure fluorescent samples, where the spectral power distribution of the light is important, under certain conditions the sample itself changes the distribution. Usually fluorescent samples are viewed under daylight and therefore also viewed under (artificial) daylight for matching production. For measuring such samples the instrument should be equipped with a source with a spectral power distribution similar to that of daylight.

Incandescent lamps or, as substitutes for daylight, xenon flash lamps (filtered or unfiltered), are used as light sources in the spectrophotometers.

Certain instruments use filtered incandescent light, sometimes in combination with an UV lamp, as a substitute for daylight. Such instruments measure fluorescent samples incorrectly because incandescent light has to be filtered so much to get a distribution similar to daylight that 99% of its radiation has to be absorbed. In this case the amount of light that falls on the detector is so small that the results are not reproducible. In addition the glass envelope (of the lamp) absorbs the UV light from 300 to about 320 nm, so that part of the UV light of daylight is not available in incandescent light. Even if the incandescent light is filtered only enough so that the distribution in the visible spectrum is equal to that of daylight and then the light of an UV lamp is added, the distribution still does not agree with that of daylight.

4.1.2. Splitting the Light into Its Component Wavelengths

The reflectance factor of a sample is measured wavelength by wavelength. To do this the light must be broken down into the individual wavelengths. It can be done before the light falls on the sample or after the reflection from the sample. Because many more samples are fluorescent than is generally thought and because some samples emit IR fluorescent light, which cannot be detected before the samples are measured with simple methods of examination, for color measurement, in all cases, an instrument should be used where the reflected light from the sample is made monochromatic. Even the use of an incandescent lamp as a source gives a measuring error smaller than that resulting from illuminating the sample with monochromatic light.

The classic element for dividing white light into its components is the *prism*. In a prism the blue light is refracted more than the red light because the refractive index of glass is a function of the wavelength. When white light passes through a prism, it is divided into the single wavelengths. With the help of a mask, the single wavelengths then can be isolated for illuminating the sample (in the case of monochromatic illumination) or for detecting the individual wavelengths of light reflected from the sample (in the case of polychromatic illumination). Because of the energy required by the detector for the detection of the light, the slit in the mask can not be made too small. Prism instruments and also all other instruments therefore measure not only the light with the indicated wavelength but also light with longer or shorter wavelengths. The distribution of the transmission corresponds with the shape of a triangle. The half bandwidth (by definition the radiation with longer or shorter wavelengths with an energy of 50% of the maximum transmission) for color measurement instruments on the market is between 2 nm and 20 nm. This is true also when the components discussed below are used. Because the reflectance curves and also the transmittance curve normally do not display abrupt changes is shape, the half bandwidth of color measurement instruments is not an important characteristic of spectrophotometers, especially when color difference measurements are made.

In modern instruments a diffraction grating is often used to break down the light instead of a prism. With a diffraction grating the light is divided by diffraction at many openings, which are placed close together, into the individual wavelengths. The diffraction grating has advantages for the producer of instruments because the distance between the individual wavelengths is constant over the whole spectrum. This is not true for a prism. The mask for getting monochromatic light therefore can be fixed and does not have to be variable as with prism instruments to get light with the same half bandwidth over the whole spectrum.

A third method for obtaining monochromatic light, the use of interference filters, has been widely used.

4.1.3. Measuring Geometry

If we judge samples visually, or stating it in another way, if we match samples visually, we do that often in daylight. The samples to be matched are laid down, such as on the window sill of a north-facing window, and are diffusely illuminated; that means with light coming from all directions (with the whole light coming from the sky). Only the light that is reflected in one direction and striking the eye is matched. If we do not match under daylight but in a light booth, the relationship between the light source, the sample, and the eye is similar. If we measure a sample, it has to be put into the light beam of the instrument so that the relationship is similar to that of the visual matching conditions. The instrument should view the sample as it is seen by the eye. Most instruments therefore have as component a sphere. The inner wall of that sphere is painted white. The sphere has openings that the light which illuminates the sample can go into the sphere and the reflected light can go out. The light source is so arranged that it is inside the sphere or is at least at the edge of the sphere, because in that case the sample is illuminated with the light that is made diffuse by the sphere. In addition the sphere has one or two more openings for the sample and the white standard. The openings for the sample differ from instrument to instrument. For most of the modern instruments they can be changed in discrete increments. The usual sample openings have a diameter from about 1 inch. Openings <0.2 inch are seldom used because of the amount of light that is necessary for measurement. They are available only, if at all, as accessories. Sample openings >2 inches are unusual depending on the instrument. Because the sample to be measured is always more or less nonuniform, as will be discussed in detail in Chapter 6, the opening for the sample should always be chosen as large as possible. The unevenness in this case influences the result less than measuring with a small opening.

Depending on whether sample and white standard are measured at the same time or one after the other, we call the instruments *single-beam* or *double-beam* instruments. Single-beam instruments are simpler in their construction but harder to calibrate. If we do not carefully and regularly calibrate such instruments, their long-term repeatability especially is worse than that of double-beam instruments. For fluorescent samples single-beam sphere instruments have a measuring error dependant on the instrument.

Most double-beam instruments on the market today also have only one opening for the sample. Part of the wall of the sphere is used as an indirect white standard. The white standard with known reflectance factors is measured only during the calibration of the instrument.

Instruments with a sphere geometry normally illuminate the sample diffusely and measure the light reflected from the sample in one direction. The reverse case is possible also, but it is not customary. Often, as standardized, the light reflected at 8° (calculated from the perpendicular of the sample (see Figure 4.1.3-2) is measured. The advantage of this is that, at another opening in the sphere, a gloss trap can be installed. In this case the diffusely illuminated sample is not illuminated at 8° and so the gloss is eliminated. Measurement with a gloss trap is correct only for samples with a high gloss. Generally we measure the reflectance factors including the gloss and eliminate the gloss by calculation. This is done by subtracting the light reflected at the surface of the sample, the gloss, from the measured reflectance factors. The size of the value to be subtracted is dependent on the air–sample refractive index. On average the correction for the gloss is 4%. The gloss can be measured with a gonio-photometer (Figure 4.1.3-1).

When we measure with a sphere, the structure of the surface of the sample is of secondary importance. This means that the measured values change only a little when samples with structure are put in the instrument in different directions. Nevertheless such samples should also be measured by presenting them always in the same direction to the instrument.

Although the sphere is painted white inside, it absorbs part of the light and therefore changes the spectral power distribution of the light source. The sphere also generally becomes dirty during use, such as when measuring textiles. That change varies with time. We get less soiling when we put the sample in from below to the instrument. This arrangement also has advantages for the handling of the samples.

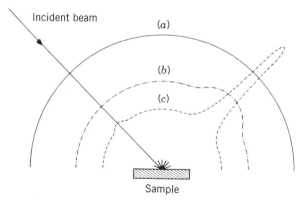

Figure 4.1.3-1. Goniophotometer curves of a sample with high gloss (c), a sample with eggshell gloss (b), and a matte sample (a). The peak of the gloss in the specular or mirror reflection of sample (c) is to be seen clearly and of sample (b) a little. (From McLaren[1]).

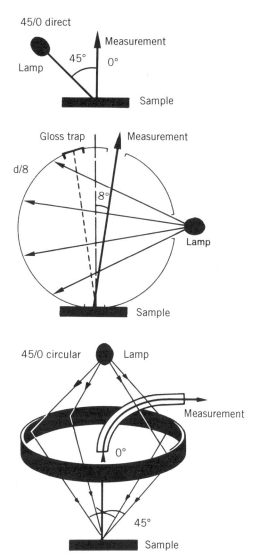

Figure 4.1.3-2. Geometries of illumination in color measurement instruments. Top: 45° illumination, 0° observation (45/0 measuring geometry). Middle: Diffuse illumination with a sphere, 8° observation (d/8 measuring geometry). Bottom: Circular illumination at 45°, 0° observation (45 circular/0 measuring geometry). (From Brockes et al.[1])

Because the modification of the spectral power distribution is especially important when measuring fluorescent samples, in some of the modern color measurement instruments the sphere is replaced by a 45° circular illumination. The reflected light is always measured at 0°, which means that the gloss is always eliminated. The circular illumination has most of the advantages but none of the disadvantages of the sphere geometry.

The spectrophotometers available today with sphere or circular measuring geometry differ in the size of the sphere and the design of the circular geometry.

For the measuring of glossy samples instruments are available in which the sample is illuminated in one direction and measured in another direction. In general the illumination comes, in contrast to the circular illumination, from only one direction at 45° to the sample. For the measurement 0° is chosen (45/0 geometry). Figure 4.1.3-1 shows clearly why the gloss in this case is eliminated. The result can be influenced a little only for samples with a very low gloss by this geometry. Such samples we should measure generally with gloss included. The gloss then should be eliminated by calculation. However, 45/0 instruments are seldom used. One reason for this is that the results of the measurements are strongly influenced by the unevenness of the surface. The reproducibility is, depending on the samples, not as good as with sphere instruments.

Schematically the three geometries discussed are shown in Figure 4.1.3-2.

Samples with metallic or pearlescent pigments ("effect" paints), which are often used in automotive paints, cannot be measured satisfactorily with the geometries described above. Recently measuring instruments (goniospectrophotometers) have become available in which the sample is illuminated directly and the reflected light is measured at several angles as shown in Figure 4.1.3-3.

The angles of illumination and observation can be changed continuously or

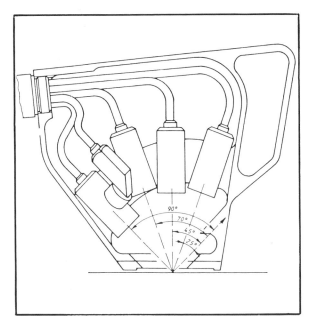

Figure 4.1.3-3. Goniophotometer measuring head, illumination at 45°, observation simultaneously at 25°, 45°, and 70° (Courtesy of Datacolor-AG).

only in discrete steps. The sample to be tested is measured in these instruments under different conditions. This type of measurement is time-consuming and at the moment there are many papers and discussions as to whether measurements at more than one angle gives a better evaluation of the sample. Recently measuring heads have been available for routine measurements where the sample is illuminated at 45° and at the same time, the light reflected at several different angles is measured. This gives naturally a much shorter measuring time (Figure 4.1.3-3). Because the measuring results of samples with "effect" paints scatter strongly, repeated measurements are absolutely necessary. Although this technique has become industrially significant in the last few years, it should be emphasized that modern "effect" paints contain not only metallic flakes but also micropearls either alone or in combination. The goniophotometeric curves, as well as the color difference that can be seen when looking at the sample from different directions, are markedly different in such cases. On the international level we try hard to establish a generally (globally) used language to describe the different illumination and observation angles as well as to establish standardized angles. For agreement probably some time is needed.

In addition there are *retroreflecting* samples, which reflect the illuminating light mostly in the direction of illumination. Street signs on interstate highways and freeways are typical examples. For the measurement of such samples special instruments are available, which will not be discussed in this book.

4.1.4. Detectors

Detectors built into the instruments are photocells, photodiodes, or photomultipliers. At earlier times the instruments had only one detector on which the monochromatic light fell and was evaluated wavelength by wavelength. To increase the speed of measurement, some of the modern instruments are equipped with 16 (or more) detectors (diode arrays). With these instruments the measurement of the reflectance curves is made simultaneously at 16 wavelengths (400–700 nm at a 20-nm interval).

4.1.5. Calibration of Instruments

The calibration of color measurement instruments today is more easily done than previously. Most mechanical adjustments are replaced by calculated corrections. Although this calibration must not be forgotten, because the results of measurements are, to a high degree, dependent on careful and regular calibrations.

The 100% and the 0% lines, and, if possible, also the wavelength and the photometric scale, must be calibrated.

For calibration of the 100% line a white standard is needed. It should be calibrated against absolute white. A white standard often used is barium sulfate pressed in a holder. The absolute reflectance factor of such a standard is about

98% for all wavelengths. The absolute reflectance factors of the powder sold for that purpose are given on the bottle. Standards of barium sulfate that are ready for use can be purchased. Barium sulfate is mostly used with real spectrophotometers that draw the curves at the same time as the measurement is made (today used almost only for research). The values given by the curves here are the relative reflectance factor against barium sulfate. If we use a plotter for drawing the curve, the measured values can be transformed, by calculation, into those measured against absolute white and the transformed values can be plotted. In such a case we can measure also against any white standard that is calibrated against absolute white.

Some spectrophotometers measure the reflectance factors only for discrete wavelengths. The plotted curve is interpolated between the measured values. For such instrument the statements in the preceding paragraph are also applicable.

To a much greater extent white standards produced from ceramics, glass, or fluorinated plastics, such as Halon, are used today. Their absolute reflectance factors are given by the producer. The reflectance factor often is lower than that of barium sulfate. Therefore they are not as good as barium sulfate for direct-recording spectrophotometers. But they are very suitable for use in modern instruments. If we use them, it is not necessary to take the same care as with barium sulfate standards, which have to be replaced often. However, the modern white standards have to be cleaned carefully prior to use.

The calibration to be done actually depends on the instrument. It has to be done in accordance with the requirements of the manufacturer of the instrument. In most instruments the absolute values of the white working standard are entered into the computer, which transforms the measured values into the absolute reflectance factors. Although the instruments used today have a good long-term stability, the calibration of the 100% line has to be done at least every day; it would be better if it were done more often.

It should be remembered that the stored values of the white standard influence the accuracy of the instrument discussed below. In spite of all the care which the calibrating institutes and the producers take in calibrating the white standards, calibration values that are erroneous and that can also change over time and lack of care in handling the white working standard, are a source of error not to be ignored in the determination of absolute tristimulus values. If we measure only color differences, this error is not important.

The calibration of the 0% line is done in a manner similar to that described above. The calibration generally is done with a black body. This is a cup lined with black velvet or a cup whose inner surfaces are painted black with a nonglossy paint. The black body has an opening of the same size as the port for the sample in the instrument. For calibration it is put at the sample opening. Because all incoming light is absorbed within the black body, its reflectance factor is 0%. The calibration of the 0% line should also be done every day. An adjustment is not necessary in modern instruments. The correction is done in the computer when the measured values of the sample are transformed by calculation with the help of the measured values of the black body. Some of the instruments on the market today are not capable of calibrating the 0% line. Purchase of such instruments is not recommended.

The described calibrations of the wavelength and the photometric scale generally are not carried out in the modern instruments. They are replaced by testing the overall calibration of the instrument (see below).

For calibration of the wavelength we use filters with sharp peaks of absorption at different wavelengths. The wavelengths of the peaks are known. Some instruments can be calibrated with low-pressure discharge lamps, which provide light with only a few characteristic wavelengths. Spectrophotometers that measure the reflectance factor for all or for many wavelengths often can calibrate the wavelength. The method is described by the producers of such instruments. With abridged spectrophotometers the changing of one or more wavelengths can be detected only by the user. The users seldom can correct it by themselves, but must ask the producer for help. Because the wavelength stability of abridged spectrophotometers is very good, frequent wavelength examination is not necessary. With real spectrophotometers the examination and calibration should be done at least once a week.

Finally also the photometric scale should be checked. Generally we do that with a set of gray filters of known transmission. The adjustment of the photometric scale has to be done as called for by the instrument manufacturers, if it is possible to do this.

All the calibrations carried out should be carefully recorded.

Especially important is the examination of the overall condition of the instrument. It is done by measuring standards with known long-term stability. The examination should be done once every week. The results are carefully examined and are carefully recorded. Long-term stable standards can be bought. (See Section 3 of Bibliography.) There are materials produced that are so stable that the standards can be taken from regular production—ceramics and enamels. Paints or plastics also can be used. Before use they have to be tested for stability. Textiles, leather and paper can not be used for this purpose. Their long-term stability is not good enough even when they are carefully stored.

For examination of the overall condition of the instrument the reflectance curves of the standards are measured at a part of the standard indicated by a mark on the back of the standard indicating the exact position for placement against the sample port. The reflectance values of the standards should be stored so that, if a large color difference is found at later date, these values, as well as the stored tristimulus values and color differences, can be checked to determine the reason for the poor stability. The tristimulus values for one or more illuminants are calculated. Later measurements are to be done with equal care. Besides the tristimulus values, color differences are also calculated. The standard for the color difference calculations are the tristimulus values of the first measurement. From the size and the kind of the color differences we can draw conclusions as to the reasons for the differences. If the color differences are much larger than the short-term reproducibility of the instrument, the calibration measurements must be carefully repeated. If the reexamination yields no better results the instrument manufacturer must be asked for help. As stated before, careful examination of the instrument should be done once every week. A slow, hardly seen, change of the measured values has an especially large influence on computer color matching. Here the measured values are compared with calibrating data that are measured a long time before the pattern that is to be matched is measured. If we measure color differences only, the change in the instrument is not so important.

4.1.6. Accuracy of Instruments

When we talk about the accuracy of instruments we distinguish between the short-term reproducibility, the long-term reproducibility, and the absolute accuracy. (In some papers often instead of *reproducibility* the word *repeatability* is used, when results from one instrument are compared. *Reproducibility* is used when results are compared that are obtained with different instruments mostly of the same type.) For description of the three quantities the color difference ΔE is used. All numbers given below are ΔE^* CIELAB values. The short-term reproducibility of an instrument is important if color differences are being measured (quality control). It is easy to test. Nobody should buy an instrument without testing its short-term reproducibility before purchase.

An instrument with a poor short-term reproducibility does not conform to today's technology. We examine the short-term reproducibility by measuring the same sample several times without moving the sample. The measured reflectance factors should not differ by more than 0.02–0.03%. With the measured reflectance factors we calculate the tristimulus values; with them the color difference ΔE is calculated. (Standard is either the average of all tristimulus values or the tristimulus values of the first measurement.) The color difference ΔE should not be larger than 0.05–0.1. Generally modern color measurement instruments meet this requirement. When the measured values show a continuous change, that may not be an error of the instrument. It can be the sample that changes. The reason for the change is the heating of the sample, especially for polychromatic illumination. There are many samples whose color changes to a greater or lesser degree on heating. A user who has to measure such samples should look for an instrument in which the samples are heated as little as possible. Suitable instruments are those which are equipped with a xenon flash lamp. Because a spot on a sample normally is measured only once, and because the modern instruments work very fast, the measuring results today are much less influenced by the heating up of the sample.

The long-term reproducibility is important for computer color matching, because the calibration dyeings of the colorants and the pattern to be matched are measured at different times. (There can be years between the measurements.) Additionally, the constancy of a production standard can be examined only with an instrument with good long-term stability.

The long-term stability as above is examined with standards known to possess long-term stability. The color difference against the first measurement should not be larger than $\Delta E = 1$, when computer color matching is to be done with success. The long-term stability of instruments that are well maintained is about $\Delta E = 0.5$. The significance, the importance and the control of the long-term reproducibility cannot be emphasized too much. In computer color matching the number of corrections can be reduced and reliable statistical quality assurance is possible when the measured results depend only on the production and not on the instrument. The long-term stability of instruments that are carefully maintained is better than the stability of many production standards. Therefore it is possible to check the stability of such standards by color measurement and to change them for new ones when they have changed. There are cases where a real production standard is replaced by its carefully measured values.

The absolute accuracy of a color measuring instrument is by far not as good as its reproducibility. The examination of it is possible only with calibrated standards. We get a feeling for the deviations when we measure samples with good long-term stability with several instruments. Years ago the results of such examinations were given in published papers. Examination of the absolute accuracy of instruments are also made in standardization committees before setting standards. Also companies that have several color measurement systems compare the results of the different systems. From all these evaluations it is known that even instruments of the same type can still show color differences of several units. The amount depends on the kind of the sample (the reflectance curve). The darker and the more brilliant the samples are, the larger the differences. If we compare instruments of different types, we see that variation increases. A further large reduction of the absolute errors of color measurement instruments is not to be expected in a short time.

In the United States a testing program for color measuring instruments is available (see Section 3 of Bibliography). The partners in that program get three samples with small color differences four times a year. One pair is nonmetameric, and the other is metameric. The participants get the results of all instruments. The scattering of the absolute values is large; the scattering of the color differences as expected is smaller than the scattering of the absolute values. But it is larger than one would wish.

The conclusion to be drawn from these facts is that it is not possible to do color matchings that are good enough when, instead of a real sample, only the reflectance factors or the tristimulus values are given. This is true also for the traffic colors described in standards. Producers of road signs do well when they ask their customers for a color standard for color matching.

A warning that has nothing to do with the absolute accuracy of instruments must be given here. Generally it is also not possible to agree on a sample from a color atlas (a color collection) for color matching. There are many reasons why it cannot be reliably assumed that samples with the same reference number in different collections are really the same. Depending on the collection and the storage conditions of such a collection, the differences can be larger than the tolerance of delivery.

Because of the accuracy of the measurement and the accuracy of the samples to be measured, the calculated results should not be printed to too many decimal places. Today it is usual to print the reflectance factors (in percent), the tristimulus values, the coordinates a^*, b^*, and L^* as well as the color difference to two decimal places. CIE chromaticity coordinates are given to four decimal places. The last digit in every case is unreliable. Printouts with more decimal places present a false accuracy that does not exist.

It is also desirable to update the software to current standards. At least two of the color measuring systems print out the metamerism index for a fluorescent lamp CWF (cool white). If we ask the user what the spectral power distribution or the correlated color temperature is for that lamp, we get no answer. In both cases the spectral power distribution is equal to that of the CIE illuminant F2.

It is self-evident that the user of an instrument knows (or should know) what observer and what color difference formula has been used for the calculations, because generally it can be selected by the user. Nevertheless it would be good if the observer and the color difference formula were always to be found along with the printed values. Today that is not always the case.

4.2. COLORIMETERS (THREE-FILTER COLOR MEASURING INSTRUMENTS)

Three-filter instruments (also called *colorimeters*) were the first color measuring instruments used in significant numbers in industry. Their name tells us about the principles of measurement in these instruments. Three-filter instruments attempt to copy the visual process in the eye, as described in Chapter 1. To give standardized results they need as light source a standard illuminant (the transformation of the light source to a standard illuminant is done with filters) that illuminates the sample as described for the spectrophotometers. The sensitivity of the detector must be transformed into the sensitivity of a standard observer also with filters. Generally one set, but a really complicated set, of filters is used for both purposes.

Generally the instruments have only one detector before which are placed, successively, three or, better, four filters. [Four filters are needed to transform the short-wavelength part of the $\bar{x}(\lambda)$ curve better than is possible with only three filters.] The values read from a properly calibrated instrument are easy to transform into the tristimulus values. The advantage of these instruments was that their measuring time was short and their short-term reproducibility was excellent. This was before digital computers were available and the modern instruments with their short measuring time did not exist. Their disadvantage is their bad absolute accuracy. The reason for this is the difficulty of reproducing the sensitivity of the eye with the combination of light source–filter–detector. For color difference measurements the poor absolute accuracy results in an error of secondary significance. The absolute errors of some instruments of this type are so large that they should be used only to measure pairs of samples with a small amount of metamerism and also with small color differences. Generally this situation exists for the quality control of products. Therefore some older three-filter instruments can be used for that purpose with good success but only in the absence of metamerism (which cannot be detected by these instruments).

The instruments sold today as colorimeters are often spectrophotometers. They are constructed like the spectrophotometers described before with some modifications. The computer installed in them is normally small. It can calculate the tristimulus values, color differences, and sometimes the metamerism index. The printout of the reflectance factors is not provided in these instruments. Because these colorimeters are simplified spectrophotometers, all statements made above for spectrophotometers apply also for these colorimeters.

Because of current state-of-the-art electronics, components, and fiber optics, it is possible to build small and transportable three-filter instruments. They all have the disadvantages described above, but with them it is possible to measure objects at places where color measurement was not possible before. We prefer to use the small modern transportable spectrophotometers.

Further there are instruments developed or modified to measure special samples such as gems. There are also standardized instruments to classify cotton or cellulose and to fix their price based on the measured values. We will not discuss these special instruments in detail.

4.3. INFLUENCE OF INSTRUMENT VARIABLES ON ACCURACY OF COLOR DIFFERENCE MEASUREMENT

More than once we have emphasized that errors in color measurement when they are not caused by the sample have only a small influence on color differences. A few examples can demonstrate this. The influence of the measurement interval, the influence of gloss, and the absolute accuracy is discussed.

Strocka[2] showed in 1973 that color differences for highly saturated samples could be measured without error with an instrument that measured only at every 20 nm. This shall be proved here again with significant examples. Therefore we again calculate the color differences for the metameric samples from Chapter 3 with a wavelength interval 20 nm. (The calculations in Chapter 3 were done with a wavelength interval of 10 nm.) Table 4.3-1 shows the color differences for the intervals of 10 and 20 nm. We see that the color difference changes only a little even for these samples with very different reflectance curves. The change is much smaller than that caused by a small change of illuminant or observer. These calculations show that not only the color differences but also the tristimulus values have not changed. Recently, abridged spectrophotometers with a wavelength interval of 10 nm have become available.

Table 4.3-1. Influence of Wavelength Interval on Color Difference; Measurement of Metameric Pair of Samples (D65/10)

Pair of Samples	ΔE CMC $(2:1)$	
	10-nm Interval	20-nm Interval
D & H F/7	0.18	0.20
CIE pairs of samples for testing artificial daylight		
1	0.04	0.03
2	0.11	0.23
3	0.03	0.16
4	0.05	0.17
5	0.04	0.02
Pairs of samples from Table 13-3		
1	0.03	0.20
2	0.02	0.18
3	0.03	0.26
4	0.02	0.27
5	0.01	0.12
6	0.05	0.15
7	0.05	0.15

Glossy samples are always matched without gloss. Often the gloss is measured and subtracted by calculation. The tristimulus values after subtracting the gloss are about 4% smaller and the saturation increases. If we calculate the color differences without subtracting the gloss, we calculate with tristimulus values that are about 4% too large. Although it is shown in Chapter 2 that the transformation of the X, Y, Z color space into a uniform color space is not linear, this error has no large influence of the color difference. The color differences discussed in Table 2-5 decrease at most about 0.1 when they are calculated without subtracting the gloss. Table 4.3-2 shows the color differences for highly saturated pairs of plastic samples. Here also the changes are small. The values of a^*, b^* are also shown; their change can be clearly seen. The color strength (see Chapter 5) is also shown; it also changes only a little (the difference in color strength always gets a little bit larger when the sample is measured with gloss excluded). Nevertheless we recommend that when the glossy samples are measured with gloss included, the measurements should be corrected by the subtraction of the effect of the gloss. Some glossy samples are so highly saturated that the calculation without correcting the gloss gives results whose errors cannot be neglected. On the other hand, these examples show indirectly that incorrect values for the white standard also give no linear changes of the reflectance values and have no large influence on the color difference.

TABLE 4.3-2. Influence of Gloss Correction on Color Differences, Color Strength, and a^*, b^* Values for Plastic Samples

Pair of samples	R_{min} with[a]	a^* with[a]	a^* without[a]	b^* with[a]	b^* without[a]	Color Strength $\Sigma K/S$ (Weighted) with[a]	Color Strength $\Sigma K/S$ (Weighted) without[a]	ΔE^* CIELAB D65/10 with[a]	ΔE^* CIELAB D65/10 without[a]
Red	23.8	38.6	40.9	14.0	15.2	104.5	104.8	0.73	0.81
Yellow 1	8.3	−14.7	−15.2	70.4	75.4	96.6	95.2	1.59	1.71
Yellow 2	7.8	0.3	0.3	76.6	84.5	99.7	99.5	0.33	0.34
Blue	16.8	− 5.3	− 5.8	−36.3	−38.7	103.1	103.5	0.46	0.50

[a]Gloss.

Table 4.3-3. Influence of Color Measuring System on Accuracy of Measuring Color Strength, Color Difference, and Residual Color Difference of Samples with Small Color Differences (Quality Control)

Test Laboratory	Sample	Tristimulus Values D65/10 of the Standard			ΔE^* CIELAB	Residual Color Difference			Color Strength $\Sigma K/S$ (Weighted)
		X	Y	Z		ΔE^*	ΔC^*	ΔH^*	
1	Yellow	68.30	75.14	21.53	0.24	0.25	−0.17	−0.18	100.1
2		67.85	74.31	21.06	0.42	0.26	−0.20	−0.14	99.0
3		68.25	74.86	21.61	0.22	0.26	−0.19	−0.16	100.5
1	Blue	16.46	18.80	44.57	0.14	0.08	−0.07	0.03	99.1
2		16.52	19.06	44.06	0.24	0.15	−0.14	0.01	98.6
3		16.83	19.44	44.65	0.20	0.08	−0.08	0.00	98.7

In any case the measurement of samples with an intermediate gloss with an instrument with sphere geometry and a gloss trap is not recommended. In comparing such samples different amounts of gloss can be excluded.

Table 4.3-3 shows the color differences measured for the same samples with different color measuring systems. Each time three samples of reference and sample (see Chapter 6) are measured, six spots are measured for each sample. Instruments of two different types were used for the measurements. The software for the calculations were the same. We can see that the agreement of the results is very good, although the absolute tristimulus values (absolute accuracy) of the instruments do not agree. If we take the average of the reference on the first instrument as standard, the absolute tristimulus values differ for the reference measured on the other instruments; for instrument 2 (same type) about 0.8 CIELAB units for yellow and about 1.5 CIELAB units for blue. For instrument 3 (different type) we calculated 0.6 and 1.8 CIELAB units. The residual color difference given beneath the color difference ΔE is the color difference that is calculated after the difference in color strength between the two samples of the pair is corrected.

4.4. MEASUREMENT OF FLUORESCENT SAMPLES

Fluorescent samples, as often said before, change part of the incident light into fluorescent light. Fluorescence often is desirable as when paper or textiles are whitened to get a bluish shade. We normally prefer a bluish white as compared with a yellowish white and call it whiter. Safety colors are, when dyed with fluorescent colorants, especially noticeable. Road workers therefore wear such clothes for their own safety. In advertising posters, fluorescent colors are widely used because they are so conspicuous. Also in fashion, these brilliant colors are often used.

All fluorescent colorants are excited by the UV wavelengths of daylight. The excitation maximum and the absorption maximum are equal; colored fluorescent samples (yellow, orange, red) therefore also are excited by the visible part of the spectrum. The wavelengths of the emitted fluorescent light are longer than those absorbed. The fluorescent light is excited by a more or less broad part of the spectrum; the fluorescent light emitted consists also of light with different wavelengths. A part of the stimulation region and part of the emission region overlap for a few wavelengths. That can give unwanted effects in some products that contain fluorescent whitening agents; they get greenish instead of whiter because the absorption edge shifts at higher concentrations to longer wavelengths (see Chapter 5).

To measure fluorescent samples we must use instruments with polychromatic illumination. This is easy to understand when we make a simplified assumption that a red fluorescent sample is stimulated for fluorescence only by light with a wavelength of 500 nm. The emitted fluorescent light should have a wavelength of only 600 nm. If we measure that sample with an instrument with monochromatic illumination at a wavelength of 500 nm, the white standard reflects light of this wavelength only. The sample not only reflects the part that is not adsorbed at that wavelength (normal reflected light) but also the fluorescent light with the wavelength of 600 nm. The detector that is in the light beam after the sample cannot distinguish between the light of the two wavelengths but (depending on its sensitivity) will evaluate the light that combines both the 500 nm (normally reflected light) along with the emitted light at 600 nm as the reflectance at 500 nm. The measured value therefore is much too large. If we then measure the sample at 600 nm, the sample reflects at this wavelength part of the incident light also. It is noted by the detector as the reflectance factor at 600 nm. Because no fluorescent light is stimulated at 600 nm, it cannot be measured at that wavelength. The reflectance factor at that wavelength is too small because the fluorescent light has already been measured at 500 nm.

If we use instead of an instrument with polychromatic illumination, the light that is reflected by the sample—it can be normal reflected or fluorescent—is then split into its components by the monochromator, it is not able to differentiate the normal reflectance from the fluorescent emission. The fluorescent light therefore is measured at the correct wavelength. Figure 4.4-1 shows the reflections curves of such a red fluorescent sample measured with monochromatic and polychromatic illumination. The relations described above can be clearly seen. If we examine these curves in more detail, we see that the reflectance factor at the fluorescent maximum is greater than 100%. This is understandable, because 100% means that all illuminating light is reflected. When the fluorescent light is added to the normal reflected light, the overall reflected light may be much larger. A reflectance curve that is measured correctly and shows reflectance factors larger than 100% always indicates a flu-

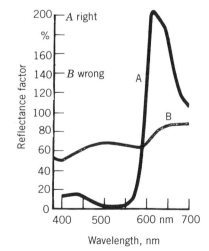

Figure 4.4-1. Reflectance curve of a red fluorescent sample with polychromatic illumination (*A*) and with monochromatic illumination (*B*) (from Brockes et al.[1]).

orescent sample. Because more samples are fluorescent than is generally thought, we again repeat the recommendation to measure only with a color measuring instrument with polychromatic illumination. If we do that, we do not have to worry whether we are measuring the sample correctly.

This statement is only partly correct. The amount of fluorescence measured is first dependent on the amount of absorbed light available for excitation. Figure 4.4-2 shows reflectance curves of a fluorescent sample measured with two different light sources. The amount of light available for the stimulation is a function of the absorption of the sample and also the amount of light available for the excitation, in other words, the spectral power distribution of the light source. If there is no light in the wavelength region of excitation, no fluorescence can be stimulated. As discussed below, the spectral power in the wavelength region of fluorescence also influences the amount of the fluorescence measured.

Fluorescent whitening agents are stimulated to fluoresce mostly by UV light, and many artificial light sources that are offered as substitutes for daylight are more or less similar to daylight in the visible range of the spectrum but have far too little energy in the UV region. They are not suitable for visual matching nor for the measurement of white fluorescent samples. (The recommendation of the CIE for testing artificial daylight lamps therefore also contains a test method for the UV region.)

For correct measurement the spectral power distribution of the incident light in the fluorescence region is significant. As stated above, the fluorescent light emitted adds to the normal reflected light. If there is much light in that range,

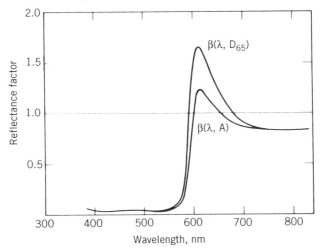

Figure 4.4-2. Reflectance curve of a red fluorescent sample measured with polychromatic illumination but with two different lamps, once with incandescent light (*A*) and once with an artificial daylight source (D65) (from Wyszecki and Stiles[1]). The reflectance curves are marked with β in the correct standardized language of the CIE and not with *R* as simplified in this book.

much light is reflected normally. Therefore the sum is affected to a high degree by the normal reflected light. If there is no light in the fluorescent region, no light can be normally reflected. We measure in that case only the fluorescent (emitted) light.

The measurement of the so called fluorescence spectrum can be used to look for the chemical structure of the colorant used. In this case we illuminate the sample only with UV light, and we measure the visible fluorescent light. The calibration of the instrument in such a case is difficult. Because such measurements are not color measurements, we will not discuss them in detail. If we need such measurements, we have to consult the literature.

In the Appendix the equations are given that describe the relations discussed on the previous page. They are given to be complete. For daily work, they are not needed.

Because fluorescent samples are viewed mostly in daylight, it is understandable after what has been said above that the light source in the measuring instrument should have a spectral power distribution similar to that of daylight. Xenon flash lamps and high-pressure xenon lamps, because their spectral power distribution is similar to that of daylight, have been found satisfactory. The knowledge of the spectral power distribution of the lamp built into the instrument really says nothing about the spectral power distribution of the light that actually illuminates the sample. Part of the light is absorbed, before it hits the sample, by optical components such as lenses and mirrors, but mostly by the sphere in the measuring instrument. If we measure fluorescent white samples with several spectrophotometers, all equipped with the same light source (a xenon flash lamp), the change of the chromaticity due to fluorescence may be twice as much in one instrument compared with another instrument. The absolute accuracy (see above) for fluorescent samples is worse than that for nonfluorescent samples.

This poorer absolute accuracy, especially for the measurement of fluorescent white samples, is due to the fact that the spectral power distribution of the source as well as the absorption of the optical system, especially the sphere, change with time. The long-term reproducibility of instruments is, for the measurement of fluorescent samples, therefore, not as good as for the measurement of nonfluorescent samples. A method was therefore developed to change the UV content of the xenon flash lamp continuously through the use of UV filters. If we use a fluorescent standard with this method, we can calibrate several instruments of the same type so that, for quality control, the instruments in several laboratories give the same values. In this case also a production sample can be used as standard; however, it is better to use a standard of known long-term stability. One producer of color measurement instruments offers measuring heads that permit this kind of UV calibration.

Instruments with sphere geometry show an additional error that we find with both single-beam and double-beam instruments. However, the error is different in both instruments. If we put a colored sample at the sample port of the sphere, the reflected light of this sample goes not only to the measuring port but also to the sphere, which is used for illumination of the sample. For two-beam instruments and nonfluorescing samples this is insignificant, because part of the sphere is used as a secondary standard. If the sample is fluorescent, the fluorescent light that goes to the walls of the sphere changes the spectral power distribution of the illuminating light. This influences the measured overall reflection. Gundlach and Mallwitz[2] show this situation clearly in a paper. Figure 4.4-3 shows the measuring results on a red fluorescent sample that supports their calculations. The measurement with the 45/0 geometry where the illuminating light is changed less than by the sphere geometry shows the highest fluorescence. If we measure with a sphere, the measured fluorescent light is smaller when the sample port is larger; this means that more reflected light falls on the walls of the sphere. Figure 4.4-4 shows the measuring results of a red fluorescent sample with different spectrophotometers. The poor absolute accuracy is clearly seen.

A further source of error, very seldom seen, may come from the total amount of light that illuminates the sample. If we use xenon flash lamps for illumination, the fluorescent power of a few samples with special fluorescent whitening agents and also of colored fluorescent samples may change, leading to erroneous measurements. The error can be reduced, but not absolutely removed, by placing a gray filter between light source and sample. The producers of instruments have worked hard and succeeded in eliminating this error.

Many published papers discuss the measurement of fluorescent samples. They describe, for example, fluorescent standards for the calibration of instruments and accessories to correct for the aging of the light source with the help of filters.

Further papers deal with the error in the calculation of the tristimulus values. As described in Chapter 1, we calculate the tristimulus values by using a

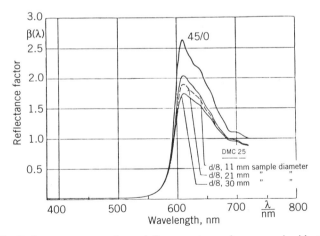

Figure 4.4-3. Reflectance curves of a red fluorescent sample measured with polychromatic illumination but with different measuring geometries. Both 45/0 geometry and sphere geometry (double-beam instrument) with variable sample port were used (from Gundlach and Mallwitz[2]).

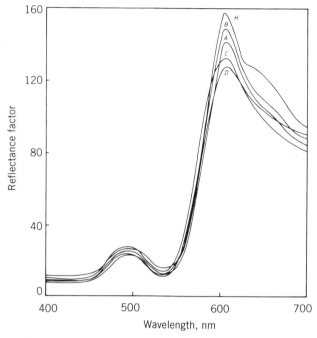

Figure 4.4-4. Reflectance curves of a red fluorescent sample measured with several spectrophotometers. All of them are equipped with an artificial daylight source and have polychromatic illumination (from McLaren[1]).

standard illuminant, a standard observer, and the reflectance curve. For nonfluorescent samples the reflectance curve is independent of the spectral power distribution of the light source in the instrument. This is not true for fluorescent samples, as is discussed in detail above. To calculate the tristimulus values that correspond with the measured reflection curves, we must calculate them with the spectral power distribution of the light that falls on the sample in the instrument. Normally we are interested only in the tristimulus values calculated with a standard illuminant. We can do that when we transform the measured reflectance factors with the formula given on p. 168 in the Appendix into those that have been measured with a standard illuminant. For that we need first the unknown reflectance curve in the wavelength range of excitation and second the unknown spectral power distribution of the illuminating light. (Such measurements and calculations can be done in only a few laboratories.) In practice we illuminate the sample with an unknown spectral power distribution, which should, theoretically, be as close as possible to standard illuminant D65 and we calculate the tristimulus values for standard illuminant D65. The absolute values therefore are not right. If the spectral power distribution of the illuminating light is not too different from standard illuminant D65, the error of the absolute values is, as described often before, of no significance for the measured color differences. As emphasized again and again for most of the questions we have to solve with color measurement, we need relative values. The quality control of fluorescent samples can therefore be done successfully in spite of all limitations mentioned above, with the help of measurements; a suitable instrument is required.

For computer color matching the fluorescence is, as described later in detail, generally ignored.

4.4.1. Whiteness

White fluorescent or nonfluorescent samples have a considerable economic significance. Since about 1950, before fluorescent optical whiteners were available, people tried hard to describe white samples with a single number for the degree of whiteness. Many formulas were suggested for whiteness. For whiteness evaluation the differences in the sensitivities of the observers and also the consumer preference for certain shades or tints are important. The preferred white is also product-dependent. Ganz[2] compiled and compared many formulas. After years of investigation the CIE decided to recommend a formula for whiteness, which is given below and in the Appendix (p. 169). If we work with a formula for whiteness, we should use this one.

CIE formula for whiteness for D65/10:

$$W = Y + 800(0.3138 - x) + 1700(0.3310 - y)$$

tint (shade)

$$T = 900(0.3138 - x) - 650(0.3310 - y) \qquad (4.4.1\text{-}1)$$

If T is positive, the sample is greenish.

If T is negative, the sample is reddish.

Limits of Application:

$$Y > 70 \qquad T < \pm 3$$

The CIE limited the use of the formula to whitenesses larger than 40. White samples with a dominant wavelength of 466 nm are called "neutral" white. Often the samples are not a neutral white but are, compared with a "neutral," reddish, or greenish. Therefore the tint of the sample should also be stated along with the whiteness.

Summary

1. A measuring instrument with the appropriate geometry suitable for the task to be solved must be selected. In most cases the ideal has polychromatic illumination with a sphere or a 45° circular geometry. If fluorescent

samples are to be measured the spectral power distribution of the light illuminating the sample should be as close as possible to the light used for visual matching—generally daylight.

2. Only instruments that are carefully and regularly calibrated will fulfill the requirements for a good instrument.

3. The short-term reproducibility of a modern color measuring instrument is so good that color differences can be measured with an accuracy that is much better than the accuracy of sample preparation and also better than the sensitivity of the eye.

4. The long-term reproducibility of carefully calibrated color measuring instruments is so good that the long-term reliability of the instruments is often better than the long-term stability of the production standards. The instruments therefore can be used for the examination of the standards. In special cases the stored results of a measured standard can be used instead of a real standard. The long-term reproducibility of the instruments must be carefully tested because, especially for computer color matching, stored data from the calibration dyeings are used. These results have been obtained long before the samples to be matched are measured. The stored data are used in the computer color matching program.

5. The absolute accuracy of color measurement instruments is much smaller. Therefore, it is not possible to do computer color matching for which, instead of a real sample to be matched, measured results from another instrument are used as standard. It is, however, good enough that color differences measured by two instruments with the same measuring geometry will be the same. (The error due to the difference in absolute accuracy is of secondary significance in this case and may be ignored.)

5

Correlation between Reflectance (Transmittance) and Colorant Concentration, Examination of Colorant Strength, and Computer Color Matching

Color measurement is used, as often mentioned, to a large extent for computer color matching. With computer color matching the concentrations of the colorants needed to match a pattern (whose reflectance or transmittance curve has been measured) are determined. For computer color matching the correlation between the measured reflectance factors and the concentration of the colorants must be known.

The same correlation is used when we determine the color strength (F) with color measurement. The color strength tells us how large the difference is, in percent, of the absorption of a sample relative to the absorption of a reference. The color strength of the reference is always 100%. (For the delivery of colorants there may be exceptions agreed to between buyer and seller.) The color strength is needed for various examinations. Mostly it is used to test a new delivery of a colorant against the last delivery. (It is better to make such tests against a reference sample agreed to between seller and buyer.) We also use the color strength to describe samples that are colored with the same colorant under different conditions of dyeing or dispersion (for pigments) or color strength as a function of time of dyeing or concentration of auxiliaries.

The reciprocal relationship ($1/F$ = dyeing equivalent) gives the number of parts (of the sample under test) that have to be used to get a dyeing that has the same depth of shade as a dyeing that is colored with 100 parts of reference.

With depth of shade we describe how strong or deeply colored a sample looks visually. For the determination of fastnesses (different types of fastness) (light-fastness, fastness to weathering, bleeding fastness, etc.), this value is important because all fastnesses change with the depth of shade. Fastnesses can be compared only when they are tested at the same depth of shade. There are formulas for calculating depth of shade with the help of color measurements with an accuracy sufficient for fastness tests.

Color strength is an important factor in the buying and selling of colorants. It is needed also for statistical quality control. Not only for these reasons in this chapter, the discussion focuses largely on color strength. It is done because the correlation between reflectance and colorant concentration is easier to describe for one colorant than for a mixture of several colorants. Figures 5-1–5-6 show transmittance or reflectance curves of sets of samples. They are colored under the same conditions but with different concentrations of the colorants. For the illustration we used only red colorants. As is shown later for each example, we used a different (chemical and physical) colorant.

First the statement made in Chapter 1 (p. 16) is illustrated: the absorbance increases when the concentration of the colorant increases. This means that the reflectance or transmittance factors get smaller with increasing concentrations of the colorants. Because the curves always show red samples the wavelength of the absorption maximum (minimum of reflectance or transmittance) is always found in the same range of wavelengths. Besides that, the curves differ very much from each other as a function of the specific colorant and the substrate being colored.

We get corresponding sets of curves with all colorants. The shape of the curves and the maximum of absorption depend on the type of colorant and its hue.

First we will look at the curves—except for one, all of them are calibration dyeings for computer color matching—without theory to get an impression of how the transmittance or reflectance curve and therefore the color of the dyeing is changed by the substrate and the colorant.

Figure 5-1 shows, as do all other figures, that the maximum of absorption is at the same wavelength for all concentrations (c). (There are exceptions for

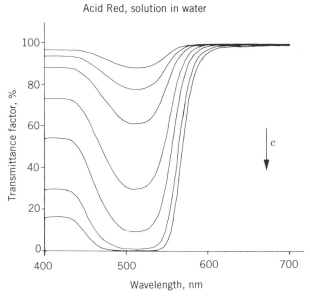

Figure 5-1. Transmittance curves of solutions in water colored with an Acid Red. Reference is the solvent (water) without the colorant. The concentration increases from top to bottom.

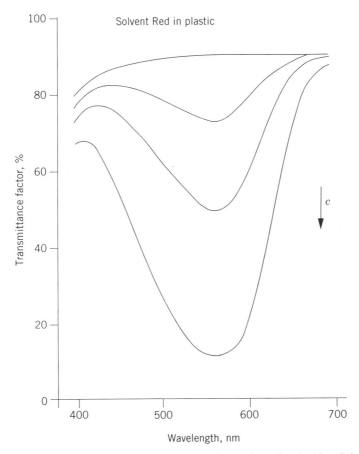

Figure 5-2. Transmittance curves of transparent plastic samples colored with a Solvent Red, measured against air. The upper curve is the transmittance curve of the plastic material without colorant.

which this is not true.) Figure 5-1 differs from all other figures in that for some wavelengths transmittance factors of 100% are measured. There are no wavelengths in the other examples with reflectance or transmittance factors larger than 97%. Figure 5-1 shows the transmittance curves of real solutions, measured against a cuvette with the same solvent without colorant.

These are solutions of a water-soluble red acid dyestuff, in the nomenclature of the *Colour Index* (C.I.[1]). The *Colour Index* is a collection of data concerning most of the commercially available (as well as many obsolete) colorants. Along with the C.I. dye class name the chemical structure is given. Colorants with the same C.I. number are usually of the same or similar chemical composition. They may differ strongly in their physical behavior. To complete the description of a colorant after the name (Acid Red) is a number, such as Acid Red 13. Under this C.I. dye class number are listed all chemically identical colorants as well as their most important properties of application

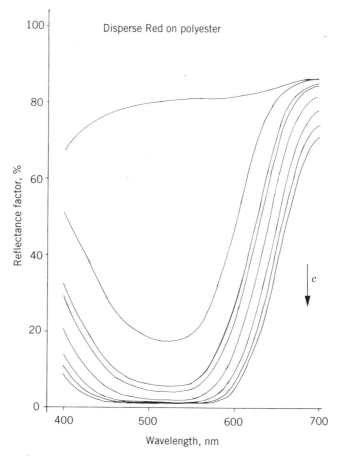

Figure 5-3. Reflectance curves of polyester dyeings, colored with a Disperse Red, measured against barium sulfate. The upper curve is the reflectance curve of a blank dyeing of the polyester material.

and their use. Where known, the five-digit C.I. number that identifies the chemical constitution is given. Under this number the chemical constitution can be found in another part of the *Colour Index*. It is unimportant for the content of this book to know the exact colorants used in the examples, but on the other hand it is important to know that the statements are correct for all colorants: the colorants discussed are identified with the name of the correct C.I. dye class involved.

Solution measurements are made to determine the color strength of soluble dyestuffs. With solution measurements we are interested only in the absorption of the dyestuff. As reference there is a "blank" of a similar solution without dyestuff.

For all other tasks (e.g., computer color matching, determination of the color strength of dyeings) we measure the reflectance factors of the samples

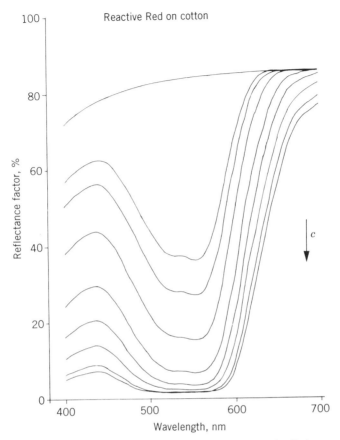

Figure 5-4. Reflectance curves of cotton dyeings, colored with a Reactive Red, measured against barium sulfate. The upper curve is the reflectance curve of a blank dyeing of the cotton substrate.

against a white standard (absolute white) and the transmittance factors against air.

Figure 5-2 shows the transmittance curves of a set of transparent plastic samples, colored with a colorant that is soluble in the plastic (solvent soluble dyestuff; Solvent Red). The curve of the plastic sample prepared in the same manner but without dyestuff is also shown. Because the measurement is done against air, the transmittance factor of about 92% is due only to a small degree by absorption of the plastic material but mostly by the reflectance at the sample–air surfaces. For computer color matching the reflectance at the surfaces is taken into account by mathematical corrections. Of interest in these curves is the apparent shifting of the absorption minimum (transmission maximum) of the colorant at about 430 nm. It is caused by the absorbance of the plastic material.

Figures 5-3 and 5-4 show the reflectance curves of textile dyeings. Figure 5-3 shows polyester dyeings with a disperse dyestuff (Disperse Red) and Figure

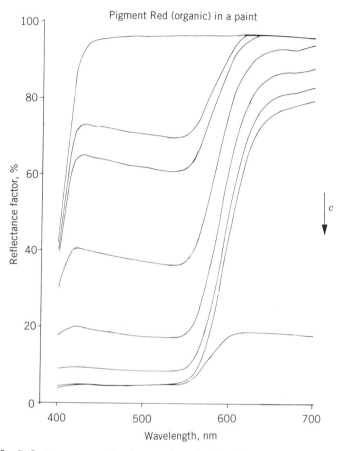

Figure 5-5. Reflectance curves of paint samples colored with an organic pigment, measured against barium sulfate. The upper curve is the curve of the paint with white only but without a colored pigment. The next five curves are the curves of the paints with white and colored pigment with the same overall concentration but with different ratios of both pigments. The second curve from the bottom is the curve of the paint with the colored pigment but without white pigment. The paint shown in the lowest curve is that of the paint with the colored pigment and with a nonscattering black pigment.

5-4, cotton dyeings with a reactive dyestuff (Reactive Red). The dyeings are measured in an opaque layer. The curve of the material used for the dyeings is also shown. It is not the curve of the original material but that of a so-called blank dyeing. This is a dyeing done under the same conditions as the others but without dyestuff. Because the material changes more or less during the dyeing process, for computer color matching we need (as will be shown in detail, p. 127) the curve of the blank dyeing and not that of the original material.

We obtain similar reflectance curves for all other textiles, paper, leather and so on.

If we look more carefully at the curves, we see that the reflectance factors

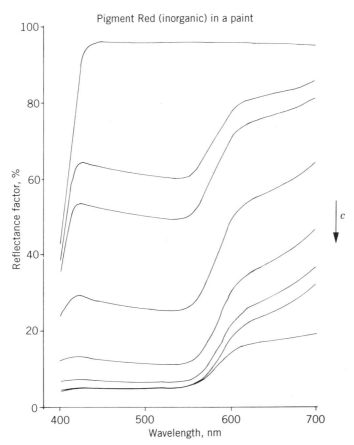

Figure 5-6. Reflectance curves of paint samples colored with an inorganic Pigment Red, measured against barium sulfate. The upper curve is the curve of the paint only with white but without a colored pigment. The next five curves are the curves of the paints with white and colored pigment with the same overall concentration but with different ratios of both pigments. The second curve from the bottom is the curve of the paint with the colored pigment but without white pigment. The paint shown in the lowest curve is that of the paint with the colored pigment and with a nonscattering black pigment.

for high concentrations in the range of the absorption maximum are the same but are not zero. This means that the absorption in this range apparently does not increase, although it can be seen that for other wavelengths more dyestuff goes to the fiber. The lowest reflectance factor in the absorption maximum is called *residual reflectance* (or *surface reflectance*). It is similar for both materials but not identical—polyester about 1.2%, cotton about 1.7%. The residual reflectance is caused by the fiber that diffusely scatters part of the incident light at its surface. For velvet the residual reflectance is smaller, for other materials it can be as large as 2%. Residual reflectance is not taken into account in the laws that describe the relationship between the reflectance factor and the concentration (see Section 5.2). For computer color matching the residual reflectance has to be taken into account. We can further see that dyeings with concentrations that are not very high in some wavelength ranges have reflectance factors that are small. These must, especially for computer color matching, be measured reproducibly and correctly. Therefore both the 100% and 0% lines must be calibrated carefully and regularly.

Figures 5-5 and 5-6 show the reflectance curves of drawdowns of paints with a Pigment Red. They are measured with gloss included, as can be seen clearly in the range of high absorption and low reflection (reflectance at the surface ~4%). For computer color matching the gloss has to be taken into account by calculation. The upper curve shows the reflectance of a drawdown of a paint prepared, as the others, but with only white pigment. The next five curves are reductions with white and different concentrations of the colored pigment (the sum of white pigment + colored pigment is constant). As long as the overall concentration of white pigment and colored pigment is not too high and as long as opaque layers are used for the measurements, such curves are approximately independent of the overall concentration. The sixth curve is the curve of the paint containing only the colored pigment measured in an

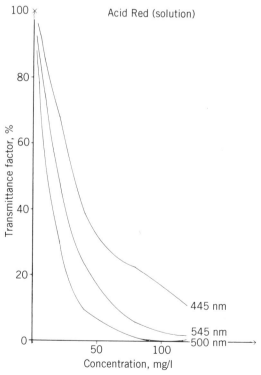

Figure 5-7. Relationship between the transmittance factor and the concentration of the colorant for three wavelengths for an Acid Red in solution in water.

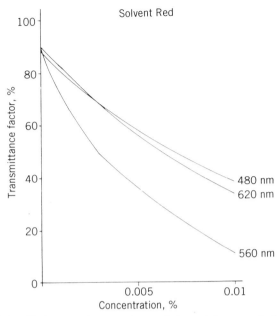

Figure 5-8. Relationship between the transmittance factor and the concentration of the colorant for three wavelengths for a Solvent Red in plastic.

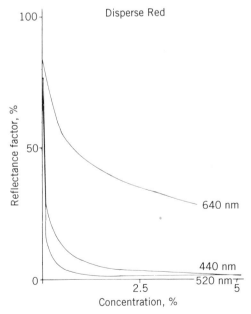

Figure 5-9. Relationship between the reflectance factor and the concentration of the colorant for three wavelengths for a Disperse Red on polyester.

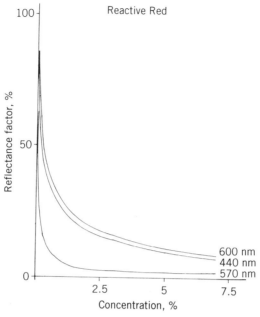

Figure 5-10. Relationship between the reflectance factor and the concentration of the colorant for three wavelengths for a Reactive Red on cotton.

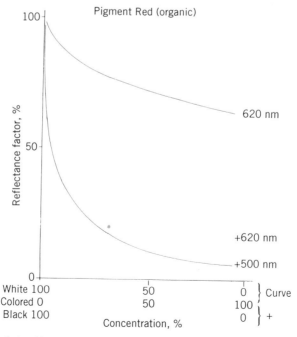

Figure 5-11. Relationship between the reflectance factor and the concentration of colorant of white, black, and colored pigments for two wavelengths for an organic pigment.

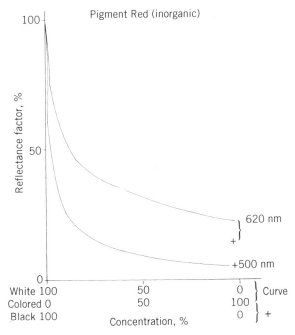

Figure 5-12. Relationship between the reflectance curves and the concentration of colorant of white, black, and colored pigments for two wavelengths for an inorganic pigment.

opaque layer. (It is the curve of the masstone.) The lowest curve is that of a paint with 98% of colored pigment and 2% of a nonscattering black pigment. (While dyestuffs generally only absorb light but do not scatter it, this is not true for pigments. For computer color matching we need, therefore, as will be shown in detail later (p. 118), more calibration colorations to determine both the absorption and the scattering values of the pigment.) In Figures 5-5 and 5-6 only the clearly seen difference between the curves in the two figures is of interest. We can see that the curve with the inorganic pigment (Figure 5-6) shows colors less brilliant than those with the organic pigment (Figure 5-5). This is caused by the differences of the absorption as well as the scattering of both pigments. Figures 5-7 to 5-12 show the relationship between the reflectance factor (transmittance factor) and the concentration for each of three wavelengths. A linear relationship is not shown in every example. For both the determination of color strength and for computer color matching we must work with functions (laws) that give at least a theoretically correct correlation between the measurement results and the concentration.

5.1. THE LAMBERT–BEER LAW

This law dates back to the eighteenth (Lambert) and the nineteenth (Beer) centuries. It states that the logarithm of $1/T$ (T = transmittance factor expressed not in percentage but as a decimal fraction between 0 and 1) for a transparent sample is proportional to the thickness d (Lambert) and the concentration c (Beer) of the colorant.

$$\log \frac{1}{T(\lambda)} = E(\lambda) = a(\lambda)cd \qquad (5.1\text{-}1)$$

Because T and R are wavelength-dependent (Chapter 1, Section 1.2), all values related to T or R are also wavelength-dependent. The term (λ) is omitted in all the following equations. The most important equations dealing with the Lambert–Beer law are given in the Appendix (p. 169).

Log $(1/T) = E$ is called *absorbance* (In Germany it is called *extinction E*. In the following equations the German letter E and not the English letter A is used.) As E is a measure of the absorption of the colorant, we sometimes also call it *absorption*.

The variable a is used here as a symbol for the characteristic of the colorant (absorption coefficient). It is wavelength-dependent and is determined therefore generally for 16 wavelengths. It applies to the unit concentration (e.g., mg/liter, %, parts). For storage of data and for testing done later with a, we must use the same unit. For most problems such as computer color matching—of mainly transparent materials—and for quality control of soluble colorants, samples are measured at the same thickness, and the thickness is often, as in the following equation, included in the constant a. The equation in that case is reduced to

$$\log \frac{1}{T} = E = ac \qquad (5.1\text{-}2)$$

Reversed, the law is given by the following equation:

$$T = 10^{-E} = 10^{-ac} \qquad (5.1\text{-}3)$$

How well the Lambert–Beer law is followed can be seen in Figures 5.1-1 and 5.1-2. The values E and c are taken from the curves of Figures 5-1 and 5-2; E was calculated from T. Both figures show a linear correlation over a broad range of concentration, as it should be, when the law is valid. If we compare the figures, we see that the curves of the solution meet at the zero point of the diagram, and those of the plastic start at low absorbance values that are wavelength-dependent. They are caused by the absorption of the substrate and increase the absorption of the colorant.

$$\log \frac{1}{T} = E = E_M + E_F \qquad (5.1\text{-}4)$$

where M stands for the material being colored (transparent plastic), and F stands for the colorant (in German, Farbmittel).

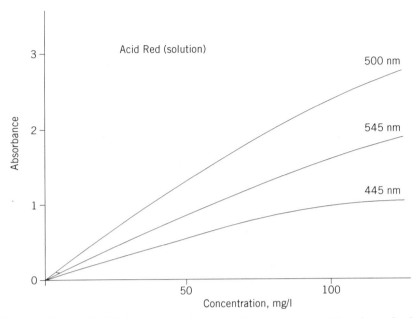

Figure 5.1-1. Relationship between absorbance and the concentration of the colorant for three wavelengths for Acid Red in water.

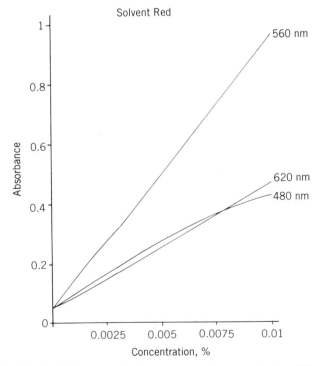

Figure 5.1-2. Relationship between the absorbance and the concentration of the colorant for three wavelengths for Solvent Red in plastic.

If we assume that the relationship given in equation 5.1-2 between the absorbance and the concentration is correct, we may write the equation as follows:

$$E = E_M + ac \tag{5.1-5}$$

For solution measurements E_M is 0, because the measurement is made against a cuvette filled with the solvent. The curves therefore start, as mentioned before, at the zero point.

With these equations it is now possible to determine the color strength by measurement. It is the ratio of the absorbance of two samples (colorations, solutions). If we designate the reference sample for the color strength with a B (B = Bezug, German for standard, type, or reference) and the sample to be tested with a P (P = Probe) when the dyeings or the solutions are at the same concentration, the equation for the color strength is

$$F = \frac{E_P - E_M}{E_B - E_M} = \frac{E_{PF}}{E_{BF}} \tag{5.1-6}$$

The number is multiplied by 100 and given in %.

For the determination of the color strength of a colorant we have to determine the ratio of the absorbance of the two colorants. That ratio shows how much more or less a shipment absorbs compared with the standard. If we replace $E - E_M$ with ac, we get the equation

$$F = \frac{a_P}{a_B} \tag{5.1-7}$$

Where the solutions being compared are at the same concentration.

When the absorption constants of the colorants are very different, we produce dyeings or solutions with different concentrations of the colorants: E_P and E_B should be similar so we avoid errors which are caused by deviation from Beer's law for large differences in the measured values. If we combine equations 5.1-5 and 5.1-7 we get the equation 5.1-8 for the color strength:

$$F = \frac{(E_P - E_M)c_B}{(E_B - E_M)c_P} \tag{5.1-8}$$

In this equation E_B and E_P are the measured and then transformed transmittance factors of standard and sample, c_B and c_P are the concentrations used, E_M is the value of the substrate or solvent without colorant. For calculation of color strength E_M often is neglected. Normally we measure at the maximum of absorption (minimum of transmission) of the colorant.

Color strength is always a relative value; its accuracy cannot be better than the standard used as reference (stability on storage, uniformity of the reference sample at different test laboratories). Because the determination of color strength, especially those by solution measurement, are only indirectly involved with color measurement, we will not discuss it in detail. The interested reader is referred to the test methods developed by the European producers of colorants. They have been published in several languages and in several journals (Brossmann et al.[2]). In that case also the paper of Commerford[2] should be read. She discusses in detail the possible errors involved with the methods. It should be pointed out that the method used for quality control of soluble colorants may be used for determination of color strength only when it has been confirmed that the results of solution measurements agree with the results of dyeing with the same samples.

The computer color matching of transparent samples is, in contrast to the determination of color strength, a problem that can be solved by color measurement. Here we shall calculate the concentrations of the colorants that we need for the matching of a sample. To solve that problem we must know the relationship between the absorbance of a single colorant and the sum of the colorants used for the match. The relationship is simple because the absorbance of the individual colorants adds up to the total absorbance. When we use, for matching of a standard with the absorbance E, three colorants (as is often the case), the mixture law is

$$E = E_M + a_1 c_1 + a_2 c_2 + a_3 c_3 \qquad (5.1\text{-}9)$$

where, a_1, a_2, and a_3 are the absorption constants of the three colorants used. They have to be determined from calibration dyeings done under the same conditions as the sample to be matched. As mentioned above, we store these numbers generally for 16 wavelengths. Because the law is, for several reasons, not valid for the whole range of concentrations, for computer color matching we have to store the data for several concentrations distributed over the range of concentrations used in application. E_M is the absorption constant of the material to be dyed, which is calculated from the transmittance curve of the uncolored material. It also has to be stored.

In equation 5.1-9 c_1, c_2, and c_3 are the unknown concentrations, which need to be determined. In the simplest case the colorants used in the pattern to be matched are known. The matching will be done with the same colorants. In that case we measure the sample to be matched at a minimum of three wavelengths, preferably at the wavelengths where the three colorants have their absorption maxima.

From the measured transmittance factors the absorbance values (E) are calculated (generally the E values were read from tables). We get three equations with three unknowns, the concentrations of the three colorants. The equations are solved with simple mathematical methods.

The case discussed above is infrequently encountered. Therefore we solve the problem with modern computers in another way. This is discussed in detail for reflecting samples below. In this case we also use the mixture law given above. The computer works with all wavelengths (generally 16) and calculates the concentrations that give the same tristimulus values for the match and the pattern to be matched. In that case we do not have to know the colorants used in the pattern.

Summary. The Lambert–Beer law describes the relationship between the transmittance and the concentration of the colorants of transparent samples.

The absortivities, which are a factor in the law, are determined with calibration dyeings. They have to be produced under the same conditions as the samples to be tested. The calibration dyeings have to be produced with several concentrations of the colorant. They must be measured at 16 wavelengths and stored in the computer.

5.2. KUBELKA–MUNK EQUATIONS

In 1931 Kubelka and Munk published equations describing the reflectance and transmittance of a translucent sample as a function of the light absorption (K) and the light scattering (S) in the sample. The absorption of the light is proportional to the concentration of the colorants used. The scattering of light, in many cases, is a function only of the material to be dyed because dyestuffs do not scatter light. If pigments are used as colorants, the scattering is the sum of the scattering of the substrate and the scattering of the pigment(s), which depends on the concentration of the pigment. The equations therefore describe also the relationship between the concentration of the colorants used and the reflectance of the sample.

The Kubelka–Munk equations are relatively complex. They are valid for diffuse illumination and diffuse reflection of the sample. Also the sample must have the refractive index of air. These conditions often are not realized for color measurement.

If the sample is opaque the equations are simplified. In general we work only with the simple equation given below. It is valid for opaque samples. Although it is valid only for the conditions described above, it has been proved satisfactory in practice.

$$\frac{K}{S} = \frac{(1 - R)^2}{2R} \tag{5.2-1}$$

where K = absorption coefficient, depending on the concentration of the colorant
S = scattering coefficient, often caused only by the substrate being dyed (If we use scattering pigments it is also caused by them depending on their concentration.)
R = reflectance factor, given in this equation not in percent but as a decimal fraction (values from 0 to 1)

For samples with a refractive index against air (paints, plastics) the equation gets more complicated. We must add correction factors.

Reversed, the equation is

$$R = 1 + K/S - \left[\left(1 + \frac{K}{S}\right)^2 - 1\right]^{1/2} \tag{5.2-2}$$

Figures 5.2-1–5.2-4 show the relationship between K/S and the concentration of the colorants, for two or three wavelengths. The numbers are taken from Figures 5-3 to 5-6, and R is transformed to K/S. If we compare Figures 5.2-1–5.2-4 with Figures 5-9–5-12 we see that the relationship of K/S against c is much more linear than the relationship of R against c. The range of linearity, compared with the transparent samples, is much smaller; for instance, it is valid for only a small range of concentrations. All curves meet the K/S axis at the point of the absorbance of the material to be dyed. The point is wavelength-dependent.

Before we discuss the limits of the law, or rather the requirements and the limits of the law for computer color matching, first we will discuss the rela-

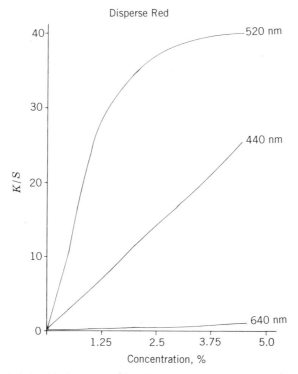

Figure 5.2-1. Relationship between K/S and the concentration of the colorant at three wavelengths for Disperse Red on polyester.

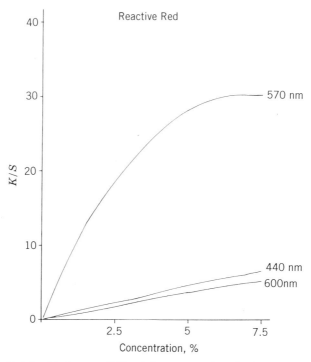

Figure 5.2-2. Relationship between K/S and the concentration of colorant at three wavelengths for Reactive Red on cotton.

tionship between K/S and the concentration of the colorants used. For reflecting samples we have to distinguish two cases: (1) that where the scattering is caused only by the substrate being dyed and (2) that where the colorant also contributes to the scattering.

Textiles, paper, and leather belong to the first group and, with limitations also oil paints and emulsion paints that are colored with organic pigments when they contain enough white pigment (that is taken as part of the substrate in computer color matching). All materials colored with inorganic pigments belong to the second group. For a nonscattering colorant the following equation is valid:

$$\frac{K}{S} = \left(\frac{K}{S}\right)_M + \left(\frac{K}{S}\right)_F = \left(\frac{K}{S}\right)_M + \left(\frac{K_F}{S}\right)c \tag{5.2-3}$$

where M = material to be colored (substrate)
F = colorant
K_F = absorption constant
c = concentration of the colorant.

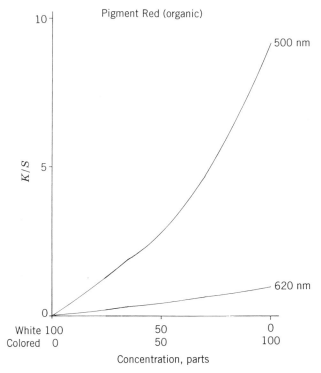

Figure 5.2-3. Relationship between K/S and concentration of the colorant reduced with a white pigment for two wavelengths for an organic pigment in a paint.

The scattering is caused by the substrate and therefore is not concentration dependent. Thus for a nonscattering colorant we store, in our computer, K_F/S as the absorption constant.

The value of K_F/S is determined from calibration dyeings. They are stored in the computer for at least 16 wavelengths in addition to $(K/S)_M$, which is determined from a blank dyeing.

For reflecting samples the absorbances of the colorants are also summed if more than one colorant is used. For three colorants the mixing law is

$$\frac{K}{S} = \left(\frac{K}{S}\right)_M + \left(\frac{K_1}{S}\right) c_1 + \left(\frac{K_2}{S}\right) c_2 + \left(\frac{K_3}{S}\right) c_3 \qquad (5.2\text{-}4)$$

If the pigment also scatters, the equation is

$$\frac{K}{S} = \frac{K_M + K_F c}{S_M + S_F c} \qquad (5.2\text{-}5)$$

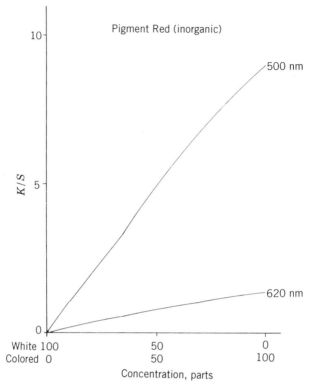

Figure 5.2-4. Correlation between K/S and concentration of the colorant (reduced with white) for two wavelengths for an inorganic pigment in a paint.

For three colorants it is

$$K_F c = K_1 c_1 + K_2 c_2 + K_3 c_3$$

and

$$S_F c = S_1 c_1 + S_2 c_2 + S_3 c_3$$

In that case the following equation is valid:

$$\frac{K}{S} = \frac{K_M + K_1 c_1 + K_2 c_2 + K_3 c_3}{S_M + S_1 c_1 + S_2 c_2 + S_3 c_3} \qquad (5.2\text{-}6)$$

where K_1, S_1, etc. are the absorption or the scattering constants of the colorants or that of the substrate. They must be determined separately.

5.2.1. Color Strength, Residual Color Difference

The application and the limits of the Kubelka–Munk (KM) equations are illustrated with the example of color strength for a nonscattering colorant. For scattering colorants the color strength due to the absorption is determined in the same manner. For such colorants the scattering power also must be determined, because that property is also important for their application.

Color strength for reflecting samples is also defined as the ratio of the KM absorption coefficients of standard and sample. Because standard and sample are dyed on the same material under the same conditions, the scattering caused by the material is the same for both dyeings.

For color strength, therefore, the following equation is valid:

$$F = \frac{K_{PF}}{K_{BF}} \qquad (5.2.1\text{-}1)$$

Using equation 5.2-3 we calculate

$$\frac{K_F}{S} = \frac{\left(\dfrac{K}{S}\right) - \left(\dfrac{K}{S}\right)_M}{c} \qquad (5.2.1\text{-}2)$$

If we put this equation into equation 5.2.1-1, we get

$$F = \frac{\left[\left(\dfrac{K}{S}\right)_P - \left(\dfrac{K}{S}\right)_M\right] c_B}{\left[\left(\left(\dfrac{K}{S}\right)_B - \left(\dfrac{K}{S}\right)_M\right] c_P\right]} \qquad (5.2.1\text{-}3)$$

The number is multiplied by 100 and expressed in percent.

The consistency and the agreement of the standard at different places of testing here has the same fundamental importance as for solution measurement.

If we describe the determination of color strength with words, we read from the equations as follows: We have to make dyeings with standard and sample of the colorant as well as a blank dyeing of the material to be dyed (the substrate) under the same conditions. (As long as the substrate is not changed the blank dyeing does not have to be prepared with every determination of color strength. It has to be dyed only once and the values have to be stored in the computer.) In the simplest case the dyeings have to be measured at the absorption maximum. The measured R values are transformed into the K/S value; or rather the R values are converted to the equivalent K/S values by a table look-up. These values are put into the equation. This simple method was used for the determination of color strength as well as for determination of the development of color strength, when used with different methods of dye odor dispersion. For that case it is still used today. That kind of determination of color strength was established in standards over the years because it was easily done; $(K/S)_M$ for that method often is neglected.

For determination of the color strength of colorant today we work with a more complicated formula.

Standard and shipment generally not only differ in a small amount in color strength but also show a small difference in shade. Since the measured color strength should agree with the color strength obtained by visual matching, we found that the color strength is more accurately determined when we calculate it not only for one wavelength but for all wavelengths (generally 16) of the visible spectrum. The color strength for the individual wavelengths have to be weighted. The formula recommended by the European producers of colorants (Baumann et al.[2]) is

$$F = \frac{\Sigma\left[(K/S)_{\mathrm{P}} - (K/S)_{\mathrm{M}}\right]\left[\bar{x}(\lambda) + \bar{y}(\lambda) + \bar{z}(\lambda)\right]c_{\mathrm{B}}}{\Sigma\left[(K/S)_{\mathrm{B}} - (K/S)_{\mathrm{M}}\right]\left[\bar{x}(\lambda) + \bar{y}(\lambda) + \bar{z}(\lambda)\right]c_{\mathrm{P}}} \qquad (5.2.1\text{-}4)$$

To determine the color strength with this formula we need a computer. Most of the color measurement systems have a program where the color strength corresponding to equation 5.2.1-4 can be calculated.

When we have calculated the color strength, we can calculate what the curve of the shipment would be if the color strength of the standard and shipment were the same. For this we first multiply the $(K/S)_{\mathrm{F}}$ values of the shipment by $1/F$ and put the corrected values in equation 5.2-3. With the corrected K/S values the corrected R values are calculated with equation 5.2-2. With these corrected R values we calculate the corrected tristimulus values and the color difference between standard and delivery corrected for color strength. That color difference is called *residual color difference*. The residual color difference is almost as important as the color strength as a quality characteristic of a delivered colorant. Unlike a simple strength difference, the residual color difference of a delivery cannot be corrected by a change in concentration.

Figures 5.2.1-1 and 5.2.1-2 are results of such examinations. They are printouts from a color measurement system of a producer of colorants. For clarity the printouts are divided into five sections marked *a* to *e*. Both printouts show the results of measurements of paper dyeings. In section *a* it is seen that both dyestuffs, standard and sample, are dyed in two concentrations. Each dyeing is measured at six different places.

(The records have been selected for special reasons discussed below; the results cannot be generalized.) The printouts, because of the many numbers printed, may seem confusing at first glance. Despite that it is important that we look at the numbers described in the text.

For now we are interested only in sections *c* and *d* of the printed numbers. The other numbers are discussed in Chapter 6. In section *c* we find the color strength calculated once at the absorption maximum and once with the reflectance factors of the whole spectrum. (The blue colorant has, for the sample delivered, a color strength different from that of the reference or standard.) In both cases the color strength calculated with the two methods differ by about 1–2%. For a delivery tolerance of 2.5% or less the method for the determination of the color strength must be agreed on. Different methods can give variations that are as large as the tolerance. In the same line, at the right side of the record, the color strength determined by solution measurement is indicated. For the blue dyestuff both color strengths differ from each other by about 10%: a proof of the warning given in Section 5.1 (p. 113). The color strength may be determined by solution measurement when it has been established that the color strength determined in solution and by dyeings are equal.

Below the color strength, in section *d* of Figures 5.2.1-1 and 5.2.1-2, the color difference between reference standard and the sample and also the residual color difference after the correction of the color strength is printed out. Experience shows that the residual color difference is usually smaller than the color difference (exceptions are possible). The data for three illuminants (TL84 corresponds to F11) and two color difference formulas are printed out. The reader may, by looking at the numbers, freshen up his knowledge of Chapters 1 to 3. In addition the metamerism index is printed out.

The errors that can be made during the determination of color strength and also by computer color matching are discussed below with a theoretical (calculated) example.

Figure 5.2.1-3 shows the reflectance curves of dyeings with two colorants for three different concentrations together with the curve of a blank dyeing. (In this example it is unbleached wool, which has a strong absorption by itself.) Both colorants have no shade difference but a difference of 10% in color strength due to the difference in concentration. The residual reflectance of the material is 2%. The difference in the curves for one concentration is hard to see, because dyeings with 10% difference in color strength show, depending on their concentration and the wavelength, a difference in the reflectance factors from 0.2 to 2%.

Table 5.2.1-1 shows the results of calculations of the color strength and the residual color difference under different conditions. The color strength is calculated once at the absorption maximum and once with formula 5.2.1-4, taking into account the whole curve. For the calculations, the absorption of the substrate is once correctly taken into account and once not. If we look only at the low and the middle concentration, we see that also the color strength at the absorption maximum is calculated incorrectly when the absorption of the substrate is not taken into account (in practice that is often the case). The error, though, is very small, but it adds to the other errors when we discuss the accuracy of the determination of color strength. If we calculate with all reflectance factors, the error increases. The lower the concentration of the colorant, the more it increases. In our case it is about 20% of what the value should be for the low concentration of colorant (107.8% against 110%). That is absolutely understandable because the influence of the absorption of the substrate increases when the absorption of the colorant decreases in comparison to the substrate. The influence of the absorption of the substrate is larger for the residual color difference. The error in this case is not insignificant. For the high colorant concentration the influence of absorption of the substrate can be ignored. Because of the influence of the residual reflection the color strength and the residual color difference are calculated incorrectly.

What conclusions can we take from these results? For both the determination of color strength as well as for computer color matching (see below) the absorption of the substrate has to be taken into account. We must always work

Sample: PAPER BLUE   Macbeth MC 2020   Xenon flash

Substrate: Paper   Dyeing concentration (%): 1.67 Sample
Corrections: SUBSTRATE   1.00 Reference

(a)

Reference (R-min.: 17.8%) ΣK/S weight.	Aver.	COV(%)	Sample ΣK/S weight.	Aver.	COV(%)	Residual DE	DC	DH	DL
11.3			12.5						
11.3			12.3						
11.2			12.4						
11.1			12.2						
11.2			12.3						
11.3	11.2	.6	12.3	12.3	.7	.46	.33	.28	.16
12.4			13.3						
12.5			13.2						
12.6			13.1						
12.7			13.1						
12.5			13.1						
12.5	12.5	.9	13.1	13.2	.6	.55	.39	.35	.17
Average	11.3			12.2					

(b)

Concentration intervals		COV(%) Uniformity intervals	
Reference:	99.3% (100.0)	110.8 (110.0)	.2	.8
Sample:	101.6 (100.0)	108.2 (110.0)	.2	1.7
Color strength sample:	109.9	104.9		

(c)

Color strength (RM)	Aim-actual (%) (:100) Uniformity intervals	COV (%)	Color Strength Solut. Measur. (%) (:100)
Aim	60.0 166.7		59.9 167.0
Absorption maximum (620 nm):	65.7 152.3 9.4		Col.Str.(SM:RM): .93
ΣK/S weight.	64.4 155.2 7.4 .3 1.8		1/(SM:RM) 1.08

Color difference:

	DE	DC	DH	DL	Da	Db	
D65/10 CIELAB	1.14	.91	.39	-.55	-.39	-.92	
CMC(2:1)	.52	.41	.24	-.21			

Residual color difference:

	DE	DC	DH	DL	Da	Db	Metamerism index
D65/10 CIELAB	.49	.36	.30	.16	-.05	-.46	
CMC(2:1)	.25	.16	.18	.06			
A/10 CIELAB	.68	.66	.12	.10	-.37	-.56	.32
CMC(2:1)	.27	.26	.06	.04			.14
TL84/10 CIELAB	.59	.52	.26	.11	-.14	-.56	.13
CMC(2:1)	.27	.22	.15	.04			.06

(d)

Tristimulus values-reference	Delta sample-reference	Chromaticity coordinates-reference	Delta sample-reference	LAB reference
X = 32.35	.178	x = .2185	-.0008	L^* = 71.28
Y = 42.59	.234	y = .2877	-.0010	a^* = -26.82
Z = 73.12	.924			b^* = -25.47

(e)

Figure 5.2.1-1. Computer record of the results of the examination of a blue paper dye with dyeings. The computer printout is divided into five sections marked a to e: a—measuring results with data for uniformity; b—accuracy of the measured data as influenced by errors in uniformity and intervals of the dyeings; c—color strength, comparison of different methods; d—color difference and residual color difference; e—tristimulus values and related numbers.

Sample: PAPER YELLOW

Macbeth MC 2020

Xenon flash

Substrate: Paper
Corrections: SUBSTRATE

Dyeing concentration (%): 2.00 Sample
2.00 Reference

Reference (R-min.: 16.7%)			Sample			Residual ΔE D65/10 CIELAB			
ΣK/S	weight. aver.	COV(%)	ΣK/S	weight aver.	COV(%)	DE	DC	DH	DL
7.6			7.6						
7.8			7.4						
7.8			7.6						
7.8			7.4						
7.6			7.5						
7.7	7.7	1.1	7.6	7.5	1.1	.20	−.18	.07	−.05
8.2			8.2						
8.1			8.2						
8.1			8.3						
8.2			8.0						
8.2			8.1						
8.1	8.2	.8	8.2	8.2	1.1	.22	−.09	.19	.08
Average 7.6			7.5						

(a)

Concentration intervals

				Uniformity intervals	COV(%)	
Reference:	101.8 (100.0)	108.0 (110.0)			.3	1.9
Sample:	100.7 (100.0)	109.3 (110.0)			.3	0.7
Color strength sample:	97.6	99.9				

(b)

Color strength (RM)

	(%)	(:100)	Aim-actual (%)	Uniformity intervals	Color Strength Solut. Measur.
					(%) (:100)
Aim	100.0	100.0			100.0 100.0
Absorption max. (620 nm):	99.6	100.4	-.4		Col.Str.(SM:RM): 1.01
ΣK/S weight	98.7	101.3	-1.3	2.1	1/(SM:RM) 0.99

(c)

Color difference:

	DE	DC	DH	DL	Da	Db
D65/10 CIELAB	.36	-.32	.16	.04	-.08	-.35
CMC(2:1)	.16	-.12	.09	.02		

(d)

Residual color difference:

	DE	DC	DH	DL	Da	Db	Metamerism index
D65/10 CIELAB	.18	-.13	.13	.01	-.09	-.16	
CMC(2:1)	.09	-.05	.07	0.00			
A/10 CIELAB	.22	-.21	.06	-0.00	-.06	-.21	.06
CMC(2:1)	.09	-.08	.04	-0.00			.03
TL84/10 CIELAB	.17	-.15	.08	-0.00	-.07	-.16	.03
CMC(2:1)	.07	-.06	.05	-0.00			.02

(e)

Tristimulus values-reference	Delta sample-reference	Chromaticity coordinates-reference	Delta sample-reference	LAB reference
X = 75.40	-.021	x = .3817	-.0004	$L^* = 94.05$
Y = 85.38	.027	y = .4322	-.0002	$a^* = -11.13$
Z = 36.77	.141			$b^* = 49.82$

Figure 5.2.1-2. Computer record of the results of the examination of a yellow paper dye with dyeings. The computer printout is divided into five sections marked a to e: a—measuring results with data for uniformity; b—accuracy of the measured data as influenced by errors in uniformity and intervals of the dyeings; c—color strength, comparison of different methods; d—color difference and residual color difference; e—tristimulus values and related numbers.

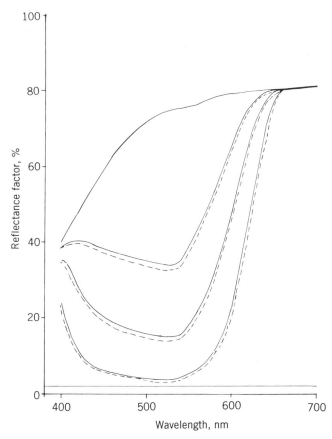

Figure 5.2.1-3. Reflectance curves of two red colorants which differ by 10% in their color strength and of the blank dyeing of the material. The curves for three different concentrations are shown.

Table 5.2.1.1. Influence of Absorption and of Residual Reflection of Substrate on Color Strength and Residual Color Difference

Concentration of the Colorant	Low		Middle		High	
Absorption of the Substrate	Yes	No	Yes	No	Yes	No
Color strength of the sample: Should be 110%						
At the absorption maximum	110.0	109.3	110.0	109.8	105.0	104.9
Weighted over all wavelengths	110.0	107.8	110.0	109.3	106.5	106.4
Res. col. diff. [CMC (2:1)]: Should be 0.0	0	0.5	0	0.2	0.2	0.1

only with the absorbance of the colorant. We get this when we subtract the absorbance of the blank dyeing from the absorbance of the dyeing with colorant.

For computer color matching, ignoring that precaution gives a larger error, because the error exists for each colorant and increases as the number of colorants increases. For computer color matching the consideration of the absorption of the substrate offers the first opportunity for correction, when we work with calibration dyeings that have been dyed on another material. Because the scattering of the substrate and therefore the absorption constants of the colorant are changed when dyeings are done on different material, the consideration of the absorption of the new material is correct but not sufficient to calculate a match with a small error.

For the determination of color strength care must be taken that the color depth of the dyeing is not too high. If it is, the result is influenced by the residual reflectance. (The residual reflectance generally is not taken into account by calculation, although formulas that include that correction can be handled by the computer without difficulty.) Among the reasons for not using such a correction is the possibility of determining color strength at depth of shades for which the influence (of the residual reflectance) is negligible. A rule of thumb says that the reflectance factor of dyeings used for determination of color strength at the absorption maximum should not be less than 5%. For computer color matching the residual reflectance is not taken into account in the formulas that describe the relationship between the reflection and the concentration (see above). The reason is that the theoretical relationship has an error not only because of the residual reflectance but also because the affinity of the dye influences the theoretical relationship. We try to take both errors into account at the same time.

Summary. The Kubelka–Munk equations describe the relationship between reflection and the concentration of the colorants of opaque reflecting samples.

The residual reflectance limits the range of validity of the equations.

The absorption constants and, if applicable, the scattering constants, which are factors in the equations, are determined by calibration dyeings at various concentrations. The calibration dyeings must be produced under the same conditions as the sample to be examined. The absorption and scattering constants are stored (generally for 16 wavelengths).

The equations for calculating the color strength and the residual color difference are described.

5.3. COMPUTER COLOR MATCHING

The possibility of doing ''computer color matching''—to calculate the concentrations of colorants that are needed to match the color of a submitted sample— with the help of color measuring systems, led to the wide use of color measurement in industry. Today several thousand color measuring systems are used for that purpose. Because measurement systems continue to improve in both cost and ease of use, the number will increase rapidly. The advantage of computer color matching against visual color matching is indisputable. This is true despite the limits of accuracy described below. The discussions of the possible errors will help raise the accuracy as high as possible. The decrease in avoidable errors contributes to the most efficient use of these techniques.

The Lambert–Beer law and the Kubelka–Munk equations have been used since the beginning of computer color matching (about 1950). We started with graphical methods. Later we used analog computers. The measuring data from a spectrophotometer had to be stored by hand in such computers. About 1970 the first color measuring systems equipped with large (and expensive) computers became available. Despite the price of at least $80,000, many systems were installed. Today a color measuring system often consists of an abridged spectrophotometer coupled with a personal computer (PC). The price of a system large enough so that computer color matching can be done is about $50,000.

From the beginning we could estimate what cost savings and what improvement of quality was possible with computer color matching. The estimates showed a cost advantage by using these techniques; that was the reason for their widespread use. As a consequence of this development, there are fewer colorists who are able to make visual color matches. For that old conventional task of color matching, a great deal of experience is necessary. Further, experienced colorists cannot compare with the speed of a computer-assisted system to get the cheapest and/or the formula with the smallest metamerism, with which a sample can be matched. Colorists need for a match, especially when they work with several combinations of colorants, a reasonable number of laboratory dyeings, which cost time and money.

Visual or computed calculated formulas must be generally corrected once or more because the first formula gives a more or less unsuitable match. Here also it has been found that the number of corrections with a color measurement system is smaller than that for visual corrections. Apart from the fact that from the computed formulas a choice can be made to select one that shows the least metamerism or the required fastnesses at the lowest cost, a further advantage is that obtained by minimizing the number of laboratory dyeings that must be made. This is a cost advantage not only in labor costs but also in saving time. (An offer can be made sooner; an order can be delivered faster.)

To be in the position to do computer color matching, we need the absorption and the scattering coefficients (the calibration values) of the colorants and the dyeing materials used. They are factors in the Lambert–Beer or Kubelka–Munk equations, as described in Sections 5.1 and 5.2. Without knowing these values, we cannot do computer color matching. The determination of the calibration constants is the first step which has to be done.

The principle for calculating the first formula with computer color matching is the same for all systems. The reflectance curve of the sample to be matched has to be measured and stored in the computer. This measurement must be made at complete hiding (opaque)—or for transparent samples, the transmittance curve must be run. For translucent patterns or for prints on paper the measurements have to be made as required by the calculation program. For high-gloss samples the program often allows the choice of measurement with or without gloss included.

Fed in to the computer after that are the data for the colorants with which the sample is to be matched and the material (paint, plastic, fiber, or fabric) and the dyeing conditions. The choice of suitable colorants is a task that still must be done by colorists, because only they know which colorants are suitable to fulfill the use properties that are asked for. Only they have the knowledge of the properties of colorants. The number of colorants from which to choose is usually made by the colorist. A reasonable number is 10. With 10 colorants, 120 combinations with three colorants or 210 combinations with 4 colorants have to be calculated. For 20 colorants the number of calculations is raised to 1140 or 4845 possible combinations. For the first case a modern computer needs about 2 minutes when the measurements are done with 16 reflectance factors. If measurements are made at 10 nm intervals, the calculation time increases. On the other hand, the speed of computers is increasing very rapidly so that such computer programs will be used more in the near future. If the result with the 10 colorants is not good enough, it is more reasonable to change some of them, and to recalculate than to increase the number to select from.

The data are sufficient for the computer to calculate a formula that gives the same tristimulus values as the sample to be matched for one illuminant-observer condition within a given tolerance (often the permitted tolerance is 0.1).

The calculation starts with estimated concentrations. (The different programs use different techniques for that first step. This may have a small influence on the time for calculation; otherwise it is unimportant.) The tristimulus values are calculated with the estimated concentrations. Afterward the concentrations are changed. Again the tristimulus values are calculated. With these repeated calculations we get the relationship between the changing of the concentration and the changing of the tristimulus values. With that knowledge it is now possible to change the concentrations more specifically so that the calculated concentrations match the pattern. For the calculation the calibration dyeings (see below) that have the nearest concentration to the calculated concentrations are always used so that deviations from theory can be evaluated. Correction factors determined from dyeings with known concentrations are also evaluated (see p. 133).

If the computer calculates a match for the submitted sample for the chosen illumination–observer condition, the metamerism indices for other illumination–observer conditions can be calculated. If the costs of the colorants are stored in the computer, the costs for the match can also be calculated. Generally we may choose how many of the calculated formulas should be printed out. Often the less expensive formulas as well as the formulas with the lowest metamerism indices are printed out. From that the colorist may choose the formula that is, from personal knowledge of the colorants and from the requirements of the customer, the best for matching the pattern. Often because of costs a greater or smaller amount of metamerism is allowed. That should be avoided because as shown in detail in Chapter 3 (p. 169), such matches are difficult to compare visually. There may be problems at the correction stage, at the delivery, and when the metameric match is used in combination with other products with a different degree of metamerism. These problems may bear no relationship to the saving in the cost of colorants.

On the other hand, cases are known where only matches with a more or less large metamerism are possible. That may, for example, be the case when the pattern and the match are made of different materials—an artist's design against a paint, a textile to match leather. In such cases it may be possible to use only colorants that have reflectance curves that deviate strongly from those in the sample to be matched. A further reason requiring metameric matches is the limited number of colorants a dyehouse will work with. The inclusion of new colorants in each case demands making new calibration dyeings; this requires time and money. The metamerism acceptable should be part of the contract between the customer and the dyehouse.

5.3.1. Determination of the Calibration Constants

The statement that computer color matching produces especially good results when the calibrations dyeings used for the calculation are done under the same conditions used for the coloring of the sample is self-evident. That this requirement never can be met exactly is one reason why the sample dyed with the calculated formulas always shows a more or less large error. To determine the absorption and the scattering constants of the colorants, calibration dyeings have to be made under the conditions that will be used to match the pattern. By "conditions" we mean the material (kind of textile material, combination of the pulp, recipe of the paint or the plastic) as well as the dyeing conditions (with that we mean also the method and degree of dispersion used in the production of paints and plastics). Because the theory that describes the relationship between reflection and the concentration of the colorant never is fulfilled exactly—residual reflectance of the material, percentage of the dyestuff on the fiber, which is dependent on the concentration, exhaustion rate, agglomerates of pigments at higher concentrations—it is not enough to prepare only one calibration dyeing at one concentration. It is necessary to make dyeings at several concentrations (calibration sets). The number of dyeings necessary is a function of the dyestuff and the dyeing conditions. The reader probably has seen that the curves discussed in Figures 5-2 to 5-6 contain a different number of samples. On average we work today with seven concentrations within a calibration set.

Theoretically we need calibration dyeings for each material and each dyeing condition, a demand rarely met in practice. Calibration dyeings that have not been optimized therefore are the largest source of error in calculating the first formula. While it is not possible to make calibration dyeings for every condition, it is, however, not sufficient to make only one set of calibration dyeings for each colorant. Dyestuffs used for exhaust dyeings, for cold-pad dyeings, and for printing each need a separate set of calibration dyeings under each of those conditions. The same is true for colorants used in different paints or plastics (composition of the recipe, condition for dispersion, etc.).

The sets of calibration dyeings must be prepared in the laboratory with the highest possible accuracy. They are measured with a spectrophotometer. Generally the measurement values are stored directly in the computer. But the measured values are tested before storing them for reasonableness (dyeing errors). The computer calculates the absorption constants and, if the colorants scatter light, the scattering constants for each of the dyeings. Theoretically these constants must be identical for all dyeings of a calibration set for one wavelength. Because they are wavelength-dependent, they should differ only from wavelength to wavelength. Figure 5.3.1-1 shows the absorption constants for the red disperse dyestuff for the same wavelengths used in previous figures as a function of concentration. They are by far not identical. In our case because

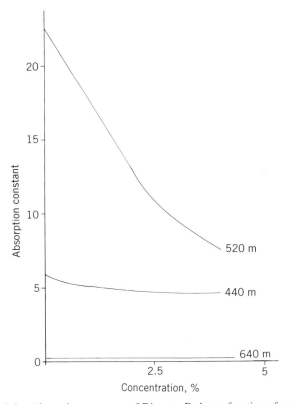

Figure 5.3.1-1. Absorption constants of Disperse Red as a function of concentration.

the dyestuff has a strong absorption and is, despite that, used in high concentrations, the influence of the residual reflectance is large. The rate of exhaustion of the dyestuff, which depends on the concentration, also has an influence, although a smaller one.

The tested and approved absorption and scattering constants are stored in the computer for a long time, often forever. The computer makes the calculations with those values that are valid for the concentrations calculated by the computer. How that is done is a function of the software supplied by the instrument manufacturer.

Together with the K and S values of the colorant, the price of the colorant is often also stored. So the price of the colorants for the calculated formula can be estimated.

The production of the calibration dyeings is work that costs time and money. Users of color measuring systems therefore try to do as little work as possible. They prepare calibration dyeing under only a small number of conditions, or they even work with the calibration dyeings (or the measured data) that they get from the producer of colorants. These calibration dyeings very seldom are done under the customer's conditions. All producers of colorants use, for the dyeing of their calibration dyeings, generally another material (kind of textile material, pulp mixture, paint or plastic recipe) and other conditions for the dyeing. When the user of a color measuring system mixes calibration dyeings from other laboratories, the calculated first formulas have a larger error than is the case under optimized conditions.

On the other hand, even under optimized conditions it is not possible to work with calibration dyeings without an error. Already the change of the material being dyed, which has not only another absorption constant but mostly also another scattering constant, has an influence on the constants stored in the computer (this applies also to the recipes for paints and plastics). The absorption of the new dyeing material can be taken into account but not the different scattering; the effect of the new substrate is accounted for by making new calibration dyeings. Textile dyers know that coarse wool yarn dyed with the same dye concentration is colored much deeper than a fine worsted dyed with the same concentration. The blank dyeings of both materials looks the same.

Figure 5.3.1-2 shows reflectance curves, measured on synthetic fibers dyed under the same conditions. The fibers differ in denier (fineness of the yarn—dtex). The difference found in color strength can be as large as 300% (Dorsch et al.[2]).

Finally we will point out a further unavoidable error. The calibration dyeings are done under optimized conditions at another time, with a different batch of material and also a different batch of colorant than those used for the match. Two samples dyed at different times under the "same" conditions (materials, colorant, dyeing conditions) are never the same. The reproducibility depends on the product. The error gets larger when we work with a new batch of material and a new batch of colorant. Deliveries of colorants vary today, usually within small tolerances, a little in color strength and shade; this difference influences

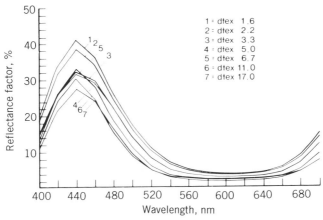

Figure 5.3.1-2. Reflectance curves of blue synthetic fibers with different fineness of yarn dyed with the same amount of a Basic Blue (from Dorsch et al.[2]).

(negatively) the accuracy of computer color matching. The error gets larger if a colorant is replaced by a colorant with the same *Colour Index* number from another source. For example, that may be the case when colorants are bought from cheap or unreliable dealers.

We must point out also that many users of colorants store open containers improperly in their color kitchens. In this way they cause a further, sometimes large, error that can be avoided. All organic dyestuffs are more or less hygroscopic (and pick up water when carelessly stored).

To minimize the errors discussed for computer color matching we can work with computer color matching with correction factors. To determine such correction factors we make one or more sample dyeings with the substrate, the colorants and the dyeing conditions to be used for the match. Dyeings with unsaturated colors are most suitable. Such dyeings we use as samples to be matched for computer color matching. It has to be done with the same dyes. Then we compare the amount of dye actually used with the calculated amount. In many cases the factor between the used and the calculated concentrations of colorants is about the same or of the same magnitude for all dyes. So an average correction factor can be determined. With that correction factor the accuracy of the first formula for matching a sample, dyed under the conditions of the correction dyeing, can be improved. Most programs for computer color matching can work with such factors. (It is also possible to work with several correction factors for each of the different influences.)

5.3.2. Calculation of the First Formula (Initial Match)

The programs for computer color matching differ in the different color measuring systems. For the determination of the calibrating constants, all of them work with the equations of Kubelka–Munk and Lambert–Beer. Also the described equations for mixtures are used. Beyond that the programs in these systems may differ. The sellers of color measuring systems try to adapt their programs to the various application conditions as much as possible. All of them offer programs for calculating formulas for textiles, paper, leather, and so on and further programs, which include the scattering of the colorant and the difference in refractive index between the sample and air. Such programs are used for the calculation of formulas for paints and plastics. Additionally, programs are offered to calculate formulas for translucent samples. For such samples the overall transmittance and the reflectance of the sample over a known background is measured. There are programs to calculate formulas for prints on paper—it can also be printed on metal, plastic, or ceramics—for the different printing techniques. Further formulas can be calculated that include more than three colorants. (For matching of paints and plastics the white pigment in most cases is an additional colorant.) The calculation of formulas for mixed fibers also is possible. The problem of calculating formulas for fluorescent patterns has not yet been satisfactorily solved. They also are very difficult to match visually. Theoretically possible programs for fluorescent samples need a very large number of calibration dyeings because the influence of the absorption of one colorant on the fluorescence of another colorant has to be taken into account. The number of calibration dyeings needed to make such a program useful is economically infeasible. For color matching of a fluorescent sample the fluorescence generally is neglected. For such samples, if possible, the colorants used in the sample to be matched should also be used for the match.

Determination of the colorants used in the pattern is possible with simple laboratory methods (chromatography, for pigments; also the comparison of spectra in solution; see Billmeyer and Saltzman[1]). That possibility, unfortunately, is very seldom used.

The programs available can be, depending on the supplier, more or less suitable for special problems. This especially applies to users who have to match samples that are not reflecting normally (this is true also for samples of paints or plastics); therefore, they should test the programs of the suppliers before buying a color measuring system.

It is not the purpose of this book to discuss the different programs. They are changed, as are all other software, more or less often. If they are not improved, at least they are made easier to use.

Because the first formula always has a more or less large error, it is recommended that dyeings be made with that formula first in the laboratory and then to correct it. Figure 5.3.2-1 shows the statistical distribution of the accuracy for calculated formulas of paints, calculated and colored under optimized conditions. For optimized conditions similar results can be found also for textile dyeings, as can be found in many papers. In MacDonald[1] the average error for the first formulas of textile dyeings with disperse dyestuffs is given as 2.0 CMC (2:1) units. Dyeings with reactive dyestuffs show an average error of about 2.5 CMC (2:1) units. We do not wonder that the accuracy for reactive dyestuffs is not as good as for disperse dyestuffs, because reactive dyestuffs show an

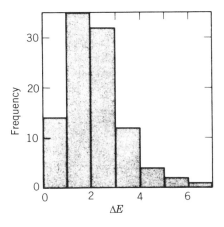

Figure 5.3.2-1. Statistical distribution of the color difference for the first formulas for matching paints under optimized conditions (from Billmeyer and Saltzman[1]). Here, ΔE is given in ANLAB units. As a rare exception, the ANLAB unit can be transformed into the CIELAB unit by a factor of 1.1.

affinity that is more dependent on the concentration than that of disperse dyestuffs. Often when the work is done under conditions that are not optimized the error of the calculated first formula may be much larger.

There are plants that work under optimized conditions and have had such good experiences with computer color matching that they transfer the calculated formula directly to their production. In that case they also have to wait for a correction where necessary. So that the correction requires only the addition of dye, they make the dyeing only with a percentage of the calculated concentration (often with 80%).

All that has been discussed above is illustrated in detail with several practical examples. Figures 5.3.2-2 to 5.3.2-12 show computer printouts from a color measurement system often used in practice. The calculated formulas are not optimized to show some practical limits for computer color matching. (Data-

Reference   :   LEMON

Yellow 4G	0.356	0.354	0.348	0.361	0.362	0.295	0.185	0.184
Gold yellow R	–	–	–	–	–	0.177	0.546	0.548
Brilliant orange 2RL	0.046	0.047	0.050	0.039	0.040	0.034	–	–
Red 5G	–	–	–	0.005	–	–	–	–
Red R	–	–	–	–	0.004	–	–	–
Brilliant blue 3RL	–	0.003	–	–	–	–	–	0.001
Brilliant blue BGE200%	–	–	0.005	0.004	0.004	0.003	–	–
Dark blue 3R	0.003	–	–	–	–	–	0.001	0.001
	1	2	3	4	5	6	7	8
Metamerism A	0.0	0.1	0.8	0.3	0.4	0.6	1.3	1.2
Metamerism TL84	0.3	0.4	0.8	1.1	1.2	0.8	0.6	0.6
Price	15.2	15.1	15.0	15.5	15.5	22.3	37.4	37.6
Color difference	0.0	0.0	0.0	0.4	0.4	1.1	0.0	0.0

Figure 5.3.2-2. Printout of calculated formulas with eight disperse dyes from which to choose for the matching of a yellow sample (D65/10, metamerism index for A/10 and TL84/10).

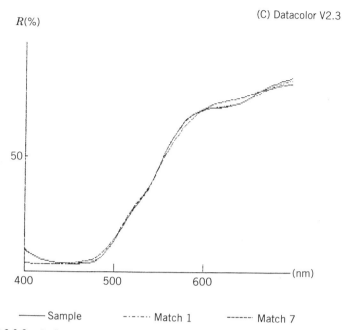

Figure 5.3.2-3. Reflectance curves of the sample to be matched as well as those of a good and a bad calculated formula (all from Figure 5.3.2-2).

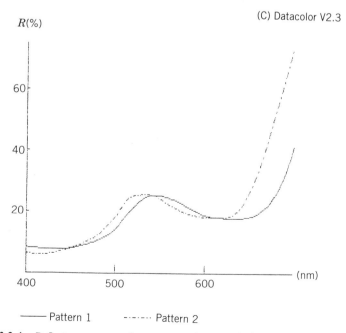

Figure 5.3.2-4. Reflectance curves of two metameric green dyeings (pattern 1 and pattern 2), which are chosen as samples to be matched for computer color matching.

Reference : Pattern 1

	1	2	3	4	5	6	7	8
Black 3RL 200%	0.118	0.118	0.118	0.118	0.118	0.118	0.118	0.071
Red R	–	0.002	0.004	–	–	0.002	–	0.013
Red 5G	0.005	0.003	–	–	0.003	–	–	–
Gold yellow R	–	–	–	–	0.049	0.051	0.113	–
Yellow 4G	0.140	0.141	0.142	0.140	0.120	0.120	0.093	0.153
Brilliant blue BGE200%	–	–	–	–	–	–	–	–
Brilliant blue 3RL	–	–	–	–	–	–	–	0.063
	1	2	3	4	5	6	7	8
Metamerism A	2.3	2.3	2.4	2.6	2.4	2.5	2.8	2.4
Metamerism TL84	1.6	1.7	1.8	1.9	1.7	1.8	1.7	2.6
Price	7.2	7.3	7.3	7.0	9.0	9.1	11.4	9.8
Color difference	0.1	0.0	0.0	1.9	0.3	0.3	0.0	1.0

Reference : Pattern 1

	9	10	11	12	13	14	15	16
Black 3RL 200%	0.059	0.057	–	–	–	0.043	–	–
Red R	–	–	–	0.010	0.027	0.032	–	0.041
Red 5G	0.018	0.032	0.031	0.019	–	–	0.057	–
Gold yellow R	–	–	–	–	–	–	–	–
Yellow 4G	0.149	0.140	0.159	0.162	0.168	0.151	0.140	0.161
Brilliant blue BGE200%	–	0.157	–	–	–	0.187	0.303	0.196
Brilliant blue 3RL	0.080	–	0.159	0.158	0.156	–	–	0.052
	9	10	11	12	13	14	15	16
Metamerism A	2.3	0.6	2.3	2.4	2.6	0.5	2.2	0.5
Metamerism TL84	2.3	2.3	2.6	2.9	3.5	3.3	2.1	4.0
Price	10.3	13.3	13.2	13.3	13.4	14.8	18.9	17.2
Color difference	1.0	1.8	0.0	0.2	0.0	1.5	0.0	1.6

Reference : Pattern 1

	17	18	19	20	21	22	23	24
Black 3RL 200%	–	–	0.079	0.077	–	0.052	0.051	0.015
Red R	0.048	0.039	–	–	–	–	0.009	–
Red 5G	–	0.011	–	–	0.043	0.011	–	–
Gold yellow R	–	–	0.297	0.514	0.313	0.511	0.529	0.597
Yellow 4G	0.157	0.154	0.058	–	0.053	–	–	–
Brilliant blue BGE200%	0.294	0.296	–	0.080	0.303	0.151	0.155	–
Brilliant blue 3RL	–	–	0.052	–	–	–	–	0.135
	17	18	19	20	21	22	23	24
Metamerism A	1.6	1.7	2.7	1.2	1.7	1.1	1.1	2.8
Metamerism TL84	3.6	3.3	1.6	1.3	1.9	1.5	1.8	1.3
Price	19.1	19.0	21.4	32.1	32.0	34.6	35.7	37.7
Color difference	0.0	0.3	0.7	0.0	1.8	0.6	0.6	0.0

Figure 5.3.2-5. Calculated formulas with seven disperse dyes to match pattern 1 (D65/10, CIELAB; metamerism index for A/10 and TL84/10).

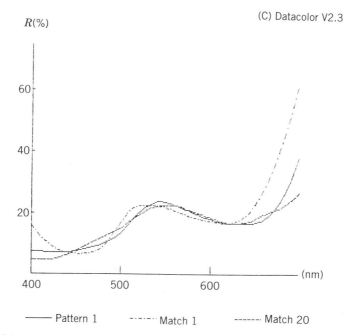

Figure 5.3.2-6. Reflectance curves of pattern 1 as well as that of a fairly good match (20) and that of a bad match (match 1).

color kindly provided the facilities for calculation in one of their sales offices.) The reader should not be troubled by all the numbers the system provides—in practice the user always has a choice of how much information is to be shown on the screen and the printouts, and should consider only those numbers discussed in the text.

All examples are calculated with standard illuminant D65 and the 10° standard observer. The metamerism index is calculated for standard illuminant A

Reference : Pattern 2

Yellow 4G	0.144	0.145	0.014	0.146	0.041	0.148	–	0.057
Gold yellow R	–	–	0.472	–	0.375	–	0.552	0.325
Red R	–	–	–	–	–	0.010	–	0.006
Red 5G	0.018	0.019	–	0.023	0.005	0.006	–	–
Navy blue 5GL200%	0.071	0.053	–	–	–	0.070	–	–
Bright blue 3RL	–	0.036	0.145	0.144	0.145	–	0.144	0.144

	1	2	3	4	5	6	7	8
Metamerism A	1.3	1.4	1.0	1.7	0.9	1.2	1.1	0.8
Metamerism TL84	1.4	1.5	0.8	1.6	0.9	1.8	0.7	1.2
Price	9.2	9.9	31.6	11.8	27.5	9.4	35.4	25.5
Color difference	0.0	0.5	0.1	0.0	1.0	0.1	1.3	1.4

Figure 5.3.2-7. Calculated formulas with six disperse dyes to match pattern 2 (D65/10, CIE-LAB; metamerism index for A/10 and TL84/10).

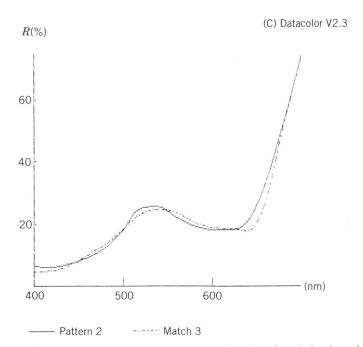

Figure 5.3.2-8. Reflectance curves of pattern 2 as well as that of a calculated match (3).

and the three-band lamp TL84 (corresponding with F11). TL84 was chosen because color matching under that source is especially critical because of its spectral power distribution. The printed color differences are in CIELAB units.

In the first example a yellow sample is to be matched on the same material. It was known that the pattern was dyed with the dyes stored in the computer; therefore the calculation should offer a nonmetameric match. For calculation we used eight dyes. Figure 5.3.2-2 shows the printout of some of the calculated formulas. The concentrations of the dyestuffs are given in percent of the weight of the material. The formula printed as No. 1 is the formula sought. Because of the errors discussed before, dyeing and measurement of the calibration dyeings and of the sample to be matched are done at different times, with new substrate and new batches of colorants. For the three-band lamp (TL84) the computed formula is not an exact match. Because of the spectral power distribution of that lamp, the smallest deviations in the reflectance curve have a large influence on the calculated tristimulus values. The printout further shows a peculiarity of the program used for the calculations. This is seen also with the further examples. The combinations printed for four colorants are calculated in a manner that the metamerism index is as small as possible. Therefore a small color difference is allowed, because the first formula has to be corrected generally, as was mentioned often before. For the selection of a formula for price and metamerism index, that small color difference is insignificant. Figure 5.3.2-3 shows the reflectance curves of formula 1 and of formula 7, which is

Reference :   Pattern 1

		1	2	3	4	5	6	7	8
FW1	Carbon black	–	–	0.006	–	–	0.007	–	0.007
GMX25	Chrome green	0.288	0.316	0.454	0.435	0.518	0.529	0.542	0.542
G920	Yellow iron oxide	0.417	0.379	0.121	0.167	–	–	–	–
AA3	Molybdate orange	–	–	–	–	0.089	–	–	–
F140	Yellow iron oxide	–	–	–	0.046	–	–	0.065	–
GLSM	Irgazin blue	0.003	–	0.006	0.010	0.011	0.004	0.009	–
BCA	Irgazin blue	0.006	0.008	–	–	–	–	–	0.003
RN57P	TiO$_2$ Rutile	0.293	0.292	0.315	0.303	0.303	0.320	0.305	0.318
FL001	Plastic	98.993	99.005	99.099	99.039	99.080	99.141	99.078	99.130
dE	D65	0.0	0.0	0.0	0.0	0.0	0.0	0.0	0.0
Price per 100		648.8	648.9	649.3	649.2	649.8	649.6	649.5	649.5
Metamerism A		0.1	0.5	0.0	0.0	0.1	0.5	0.7	1.1
Metamerism TL84		1.4	1.2	2.3	2.5	2.5	2.4	2.6	2.5

Figure 5.3.2-9. Calculated matches with seven chromatic + black and white pigments to match pattern 1 (D65/10, CIELAB, metamerism index for A/10 and TL84/10).

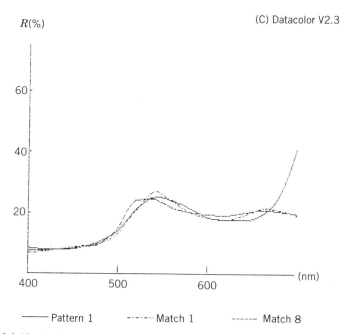

Figure 5.3.2-10. Reflectance curves of pattern 1 as well as that of a fairly good match (match 1) and that of a bad match (match 8).

Reference : Pattern 2

		1	2	3	4	5	6	7
W1	Carbon black	–	–	0.006	–	0.007	–	–
MX25	Chrome green	0.587	0.571	0.560	0.559	0.549	0.544	0.402
920	Yellow iron oxide	–	–	–	–	–	–	0.336
A3	Molybdate orange	–	0.057	–	–	–	0.079	–
140	Yellow iron oxide	0.044	–	–	0.058	–	–	–
LSM	Irgazin blue	–	–	–	0.010	0.005	0.011	–
CA	Irgazin blue	0.007	0.008	0.004	–	–	–	0.009
N57P	TiO_2 Rutile	0.328	0.324	0.337	0.327	0.341	0.328	0.339
L001	Plastic	99.033	99.040	99.094	99.047	99.099	99.038	98.915
E	D65	0.1	0.1	0.0	0.0	0.1	0.1	0.0
Price per 100		649.5	649.7	649.5	649.5	649.5	649.7	648.9
Metamerism A		1.7	2.0	2.6	3.2	3.4	3.8	3.2
Metamerism TL84		1.6	1.5	1.4	1.3	1.2	1.0	2.7

Figure 5.3.2-11. Calculated matches with six chromatic + black and white pigments to match pattern 2 (D65/10, CIELAB; metamerism index for A/10 and TL84/10).

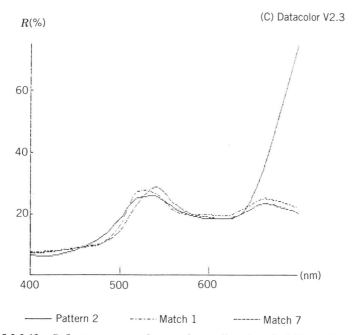

Figure 5.3.2-12. Reflectance curves of pattern 2 as well as that of a fairly good match (match 1) and that of a bad match (match 7).

more expensive and has a larger metamerism index. The deviation in the reflectance curve by an erroneous choice of colorants (formula 7) can be clearly seen.

For the next examples 2 green metameric cotton dyeings are chosen as samples to be matched (pattern 1 and pattern 2). The reason for this choice was that, based on experience, it is very hard to make nonmetameric matches with green colors, if we do not use the same colorants. The reflectance curves of the two patterns are shown in Figure 5.3.2-4. The scale of reflection in this and the following figures is expanded to 0–75% full scale, to show the differences as clearly as possible. The requirement was to match the samples once on polyester with disperse dyes and once in plastic with pigments.

Figures 5.3.2-5 to 5.3.2-12 show parts of the printout of the computer and reflectance curves of some of the formulas. Calculations were done with six to seven dyes or pigments. We will see that none of the calculated formulas is optimal.

For the textile dyeings we calculate with six to seven selected dyes as well as with all 12 stored colorants (used by the customer) in the computer. The printouts with the six to seven colorants show the formulas with the best suitable colorants for the match.

To calculate a nonmetameric match on polyester, additional dyes had to be stored and used for the calculations.

Figure 5.3.2-5 shows part of the printout of the computer. There are 24 formulas printed. We can see that it is not possible to get a formula with metamerism indices smaller than 1 for both light sources. To get a small metamerism index expensive dyestuffs Gold Yellow R and Brilliant Blue BGE 200% have to be used. Figure 5.3.2-6 shows the reflectance curves of pattern 1 and the calculated matches 1 and 20. The deviation of the curves of the formulas from that of the pattern and the influence of the colorants used can be clearly seen. Figures 5.3.2-7 and 5.3.2-8 show the corresponding calculations and curves for pattern 2. That pattern has a reflectance curve which can be a better match with the available dyes. Most of the metamerism indices are smaller, but not as small as desirable for a match. In that case we should consider dyeing with the formula which gives a low price. Figure 5.3.2-8 shows the reflectance curves of pattern 2 and that of the calculated match 3.

Figures 5.3.2-9 to 5.3.2-12 show printouts or the reflectance curves for the matching of plastic samples. If we look carefully we will see that the printouts differ a little bit from those for textiles. The printouts are adapted to the practice in the different industries. The concentrations of the pigments are not applied to the amount of plastic but so that the overall concentration (plastic + pigment) is 100. Further we can see that with the pigments chosen no large differences in the prices can be seen. In contrast to the textile dyeing, here also the price of the plastic is added in the price of the formula. Further, we can see that pattern 1 is better than pattern 2 as a match in plastics. The metamerism indices are of the same magnitude as those for the textile matches. For standard illuminant A pattern 1 can be matched with a metamerism index of 0, but not for

TL84. This means that despite the metamerism index for standard illuminant A = 0, it is not a non-metameric match; this can be clearly seen in the reflectance curves.

5.3.3. Correction of the Formulas

As has been mentioned often before, the calculated first formula often has a more or less large error. Therefore first a dyeing should be made in the laboratory with that formula. Usually the laboratory dyeing is measured and corrected by calculation. The correction depends on the color measuring system used. The corrected formula is dyed again, measured again, and corrected again. It has been shown that more than two calculated corrections mostly do not improve the match. One reason for that is the limited dyeing and measuring accuracy (a function of the sample). It can produce a corrected formula that in extreme cases can be a worse match than that with the earlier formula. Such a case is shown in Figure 5.3.3-1.

In a plant that has poor reproducibility of laboratory dyeings it is often thought that the unsatisfactory calculated formulas are the fault of the color measurement system. Management then hopes, erroneously, to get a good match faster if it corrects its dyeings visually. Production that gets unsatisfactory results after two corrections should look for the sources of error (accuracy of laboratory dyeings, unsatisfactory calibration dyeings, wrong measurements because the dyeings are nonuniform, etc.).

But there are reasons, which may require a meaningful visual correction, because the instrument has, as does the eye, difficulties to compare the pattern and the match. Such is the case when, for example, the sample to be matched

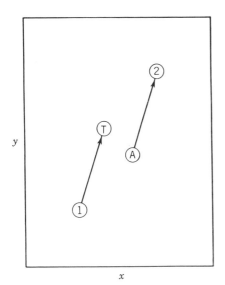

Figure 5.3.3.-1. Error due to the improper correction of the first formula of a match. The pattern (T) is to be matched. The measurement of the dyeing of the first formula based on a single sample measured once, gives the wrong chromaticity coordinates (point 1). The correct measurement, based on multiple samples and measurements, would give the chromaticity coordinates (A). Point 1 was measured and from here the correction is calculated. The calculated correction which should have moved from point 1 to the "target" T, instead produces a sample (dyeing 2) with the chromaticity coordinates (2). It has a larger difference from the target (T) than (A) (from Billmeyer and Saltzman[1]).

and the proposed match have a large difference in structure or a large difference in gloss (glossy drawdown of a paint against suede, yarn against cloth, high-gloss print against corduroy). If the pattern is very small, the measuring results sometimes also differ from the visual results (very small pattern vs. match of normal size). In that case neither the calculation with the 2° standard observer nor that with the 10° standard observer corresponds with the conditions of matching.

Some formulas, for example, contain only colorants of similar hue or only two colorants that can be corrected only with great difficulty by calculation. These are to be seen as exceptions.

Besides the correction of formulas in the laboratory, which generally is necessary to get a formula that matches the pattern with satisfactory accuracy, it is also, unfortunately, necessary to correct the formula again during the dyeing process in the plant. That can also happen when a standard formula is repeated, because not all the components (substrate, colorants, conditions of dyeing) are standardized well enough. For correction of production samples we use the same programs as for the correction of formulas in the laboratory. For that case also special programs are available, which calculate only positive corrections (additions). When necessary, the correction in that case calculates a dyeing that is a little deeper but shows the smallest possible color difference against the sample to be matched.

Whereas the accuracy of a normal calculated correction is limited by the accuracy of the dyeings, the correction of production batches is influenced by the change of the sample in production during the time of correction. That change (e.g., the exhaustion of dyestuffs between the time the sample for correction is taken and the time the correction is added) cannot always be avoided.

Summary

   Computer color matching is the task for which most of the color measuring systems in industry are used.

   For calculating formulas we need calibration dyeings. The accuracy of these limits the accuracy of the calculated formulas.

   For calculation we use the Kubelka–Munk equations or the Lambert–Beer laws. Deviations from the laws are taken into account with the stored constants of absorption and scattering.

   By calculation formulas are determined, which have the same tristimulus values as the sample to be matched for a selected illuminant–observer condition.

   The metamerism index of the calculated formula can be calculated for a specified illuminant–observer condition (or those available from the computer program). Also the price of the proposed match can be calculated.

The initial formula calculated is usually not good enough. It must be corrected.

The correction can also be calculated. The calculated corrections are limited by the accuracy of the sample preparation in the laboratory, by the relative structure and size of pattern and match, and, although seldom, also through the nature of the selected formula.

6

Influence of the Sample on the Accuracy of Color Measurements

Only at first glance is it surprising that a whole chapter of this book is dedicated to the sample to be measured. [The sample here can represent the product (as produced) to be tested as, e.g., for paper or it can be specially made to convert a liquid paint to a painted panel.] The sample to be tested generally does not fulfill the requirement of being representative of the whole lot (batch), nor can it be produced several times exactly the same in the laboratory. For measurement the sample that is presented at the sample opening of the instrument is evaluated. The result therefore can be only as good as the sample to be tested.

There are many reasons why samples are not representative. Some of them shall be discussed. They should stimulate the reader to think about how to get a measuring result that can be trusted.

The examples of this chapter are mostly drawn from the experience of colorant producers. They must work hard on this problem because deliveries of colorants are allowed to have only very small tolerances. The quality of a delivery must be guaranteed by a certificate. The accuracy for that kind of test is especially high because each time only one colorant is tested by laboratory methods that are carefully developed and standardized. The scatter of results found by these methods should give all users of color measurements cause and time to think. The customer can generally not work under such standardized conditions.

Papers about tolerances in the industries that produce colored materials are rare. The small amount of data available indicates that the tolerances are much larger than generally claimed.

That statement can easily be tested by the readers of the book because they are also consumers. If they look around their homes they will find, as I do, many products that are composed of different parts.

Nowhere do the colors match within small tolerances. At my telephone the body and the receiver has a visual color difference. The printer of my computer has at least four parts that differ significantly in hue. Wallpapers from one batch show sometimes large differences in hue from roll to roll and also in the color of the two borders. Unicolor tiles differ visually in color. I have been surprised more than once how large the differences can be, when I bought embroidery thread or knitting yarn from different dye-lots. Some clothing I have bought have some sewn-on pockets with a different color than the part they are sewed on. I own two copies of a textbook; both red covers show a visible color difference. I also have bought dresses where the different parts show a large metamerism. All that seems to scarcely disturb the normal buyer (as well as me). But it is a sign that the small tolerances the industry speaks of in reality are not as small as they claim.

Discussed below are the following sources of error: evenness of a sample and errors during the preparation of a sample. Here we have to discuss the short-term and the long-term reproducibility.

In spite of the use of very modern dyeing machines and working with the highest known accuracy in both the laboratory and increasingly in the plant, these errors persist. They cannot be avoided. In contrast to the measuring instruments, which are today much more reproducible than they were years ago, there has been no corresponding advance in the accuracy and reproducibility of dyeings done with analytical accuracy in the lab.

In the following pages we try to separate the sources of errors as much as possible. Because often several errors exist at the same time, that will not be possible in every case.

Before we start discussing the errors, the production of the many different samples that are measured today are briefly discussed.

1. Textiles.
   a. Production. Raw stock, yarn or piece goods are dyed. In a few plants the fiber is dyed during the spinning process. (That case will not be discussed further). A large number of machines are available for coloring: winches for yarn or piece goods, beams and jets for piece goods, package dyeing machines for yarn, and continuous or semicontinuous machines for pieces, to name only a few of them. In most cases a finishing process follows after the dyeing, mostly to improve the wearing characteristics. The finishing process usually alters the color. The finishing process may be also used to change the appearance of the fabric, such as to make chintz glossy by calendaring or to raise the nape of the goods with a napping machine.

   For computer color matching the pattern often is finished and the influence of the finishing has to be taken into account by calculation.
   b. Laboratory. In the laboratory we have automatically controlled dyeing machines. In them small amounts of the materials discussed above can be dyed under many conditions. The automation can be so complex that the dyeing beakers can be automatically filled very accurately. There are also small laboratory dyeing machines that can simulate continuous dyeings reasonably well. In every case the agreement between the results of the laboratory dyeings and the dyeings in the plant must be verified.

For measurement piece goods can be measured directly, although the conditions for the measurement sometimes have to be standardized (suede, knitted piece goods). Yarn and raw-stock material have to be converted (prepared) into a form that is suitable for measurement. For example, skeins are stretched on a sample holder or wound on a card. In exceptional cases yarn can be knit into piece goods. Further it can be pressed into a sleeve and cut at one edge. In practice it can be seen again and again, that yarn is measured without any preparation. Every user of an instrument who works so is advised (see p. 151) to put the "same" spot again at the instrument—and will be astonished at how large the differences in the measured numbers are. Raw-stock material generally is pressed and measured in a sample dish behind glass (see Hunter[1] and Stearns[2]).

2. Paints, printing inks, colored pastes for foamed plastics.

   a. Production. The often large number of components—solvents, resins, pigments, additives to get certain technical properties—is more or less automatically added into large tanks in a given order and then mixed. The pigments generally are separately dispersed with some of the other components before they are added. There are several machines available to do that. The degree of dispersion normally is tested before adding the pigment to the mixing tank. The viscosity of the mixture can differ very much, depending on how the material is used. It should be tested and adjusted as a part of quality assurance.

   b. Laboratory. There are machines for dispersion that are able to produce small amounts of the products. Often the method of dispersing and the method of working are only similar to those in the plant. The agreement of the results from the laboratory and the plant therefore must be carefully tested. Pigments sometimes are dispersed on a Hoover muller. The results with that machine, because of the different resins used and also because of the different conditions of dispersion, do not always match production.

The products produced as described above must first be converted into a sample suitable for measurement. (The dispersions can be tested directly in a sample dish, but the results generally do not correspond to those under application conditions.) Depending on the application we can, for example, make drawdowns, and dry them in air or in an oven. The drawdowns can be made with doctor blades or spraying, and the thickness of the layer can be very different as required by the function of the paint. The thickness of the layer and the background plays an important role in testing printing inks, which are always more or less transparent. Printing inks for textiles are printed on a small laboratory printing machine and tested as a textile print. Dispersions for coloring foam material generally are foamed in the lab. There are also more simple techniques, which must be tested to see if they give comparable results.

When testing dispersions we have to know that they can flocculate or separate during storage—it is well known to everyone who buys paints to paint their walls at home—and that they must be carefully mixed before testing them.

In connection with the testing of paints and the possible additivity of testing errors a paper by Johnston[2] must be mentioned. It is a classic among the papers about color measurement and is recommended to the readers of this book.

It is also possible to test the final products made with the dispersions described. The measurement of traffic signs was mentioned in Chapter 1. Printed papers are tested often with special instruments developed only for that purpose. Special instruments include those previously mentioned to measure artwork (paintings, murals). The final products have to be tested in place with measuring instruments specially equipped for that purpose (e.g., those with a portable measuring head). The measuring of the color of repaired cars has an important economic significance. The color of the repaired part must be the same as that on the other parts that normally are not the original color because it has changed during the use of the car. To measure them we have to realize that many of these paints are so-called effect paints that contain, in addition to normal colorants, metallic flakes and/or multilayered pearl essence flakes, which have to be tested at several illuminating and viewing conditions. For such materials the same special measurement conditions must be followed when testing the production or laboratory samples, which are described above (p. 81).

3. Plastics.

    a. Production. Many methods are used to produce colored plastics. A method often used is to mix granules of the plastic material and colorants in an extruder and to mold the colored plastic granules in an injection-molding machine. Plastics also can be produced on three roll mills or similar machines to get films of different thickness. Fluid resins can be cast.

    b. Laboratory. The machines used in the laboratory are similar to those used in production. In all cases they are expensive and large pieces of equipment.

    Generally plastic samples can be measured without preparation. The colored (pigmented) plastic granules can be tested, before use in production, on the machines in the laboratory to be sure their color is right. For such a laboratory test it is also necessary to check if the results of the tests in the laboratory and in production are the same because, for example, the temperature in both pieces of equipment may not be the same. Also it can determine if the material to be tested is homogenous, because it consists of the plastic material, a white pigment, the colored pigments, and additives. The final plastic products can also be measured. The measuring instrument chosen must be suitable for that purpose.

4. Paper.

    a. Production. Paper is produced on more or less large paper machines. Most of the papers produced are white. The colorants are added to the pulp either in the Hollander, the mixing tank, or direct to the final pulp before it gets to the screen of the machine. The paper may be coated at the sizing roll. For coated papers the top coat may also contain colorants.

b. Laboratory. Mass dyeings can be done easily in the laboratory with hand sheet formers. Also the coloring in the size roll or the coating of the surface can be done in the laboratory with simple methods, but only by using previously produced papers, which means not during the paper production process as in the plant. There are also small paper machines available on which papers can be produced on a technical scale. The cost to produce papers on such a machine is much greater than by using a hand sheet former. The transfer of the results from the laboratory to the plant has to be tested also for papers.

Paper as received can be measured with little or no preparation. Therefore it is understandable that the so-called on-line measuring instruments for continuous quality control are used mostly only in the paper industry. Although available for years, on-line measuring instruments are seldom used in industry. One reason for that is the more or less variable distance between the measuring head of the instrument and the sample to be measured. The paper (or the textile) strip always flutters a little bit. The measuring results are therefore incorrect. The newest generation of this kind of instrument measures a large part of the sample, and so the result is relatively independent of the distance between the measuring head and the sample. Perhaps these instruments will be used in a larger number in industry for continuous quality control. It is, however, difficult to solve the problem of coupling the measuring results and the required addition of colorant. At this time it seems that each case requires a unique solution. To solve that problem it is necessary to get a consistent production. Only then when the producers of color measuring systems help in every single case will more instruments of that kind be used.

5. Leather.
   a. Production. Leather is dyed, after tanning, in a separate production step or also together with the post-tanning operation. The dyed leather afterward gets one or more coatings—by doctor blade or flow coated with or without embossing. The dispersions used for the coatings are mostly colored. The coating may be more or less opaque.

   b. Laboratory. The dyeing techniques used in production can also be done with relatively simple equipment in the laboratory. The dyeing result depends greatly on the leather being used. Two dyeings done with two different parts of the same piece of leather may show a large difference in color.

Leather samples can be measured without any further preparation.

The enumeration of the samples to be prepared and measured is surely incomplete and arbitrary. But it shows clearly that there can be no general rules for the measuring of samples. A few principles are true for all samples to be measured. Each new but also experienced user of color measuring systems should do the tests described below. They are fast and simple.

First the sample to be measured should be measured several times without moving it. The color measuring systems then will use the first measurement or the average of all measurements as standard. As a result you will get very small color differences, because the color measuring instruments give, as mentioned often before, results that are very reproducible. For a few samples, measured with an instrument that illuminates the sample with the polychromatic light of an incandescent lamp, a systematic change of the measuring results may be found. Such samples are temperature-sensitive. If they are heated they change their color to a greater or lesser degree. For practical measurements that is not important, because the same part of the sample generally is not measured more than once. The measuring time is very short. You should know that it is not good to interrupt the measurement for a telephone call or a conversation. In such cases the measurement should be finished or the sample should be taken from the instrument.

After that the test should be repeated; this is the second trial. The sample should be taken from the instrument and replaced at the sample port. Every effort should be made to always measure the same part of the sample. For some samples you will find a larger scattering of the results. Knit pieces, for example, are unintentionally more or less stretched. Samples with a more or less large structure will be, again unintentionally, turned a little. Or the part measured may not be always the same. Samples with a coarse surface may be pushed a little into the sphere and not always to the same degree; therefore such samples often are measured behind glass. Much scattering indicates that a better measuring technique has to be developed first, to get results that are reproducible.

For the third test several places on the sample should be measured with the measuring technique developed before. For some samples that means a new sample has to be prepared for measuring. For molded plastic samples, for example, several samples produced with the same technique should be measured. Stretched samples of yarn must be stretched again on the holder, and yarn pressed into sleeves must be pressed again. Drawdowns also must be repeated. The new drawdowns and the old ones should be measured at several spots when that is possible. This trial will show that the samples have more or less large color differences from spot to spot or between the different samples. I know of no sample that is absolutely uniform. With this trial you will see further that the scattering gets smaller when the port of the measuring instrument is chosen as large as the sample permits, because in that case the average of a larger area is measured. You must be sure that only the sample to be measured and no part of the sample with another color is put at the sample opening of the instrument. Very small samples, such as unusual patterns or prints, should be viewed before measurement if the instrument is equipped to look at the sample port.

For the fourth trial a new sample should be prepared in the laboratory (new dyeing, new preparation of the dispersion of a paint, new preparation of the colored plastic granules). That sample should be measured as the samples on the third trial at several spots. The color differences between the average tristimulus values of the measurements of the third and the fourth trials should be compared.

We can expect the color differences to increase from trial to trial. We now should try to get smaller differences. That can be done by changing the conditions for the laboratory dyeings and for the preparation of the samples. In the plant the production conditions should be changed.

Finally the standard shall be discussed, which is necessary for all applications of color measurement. For color matching the pattern is standard. In many cases it will be of another material. From that arise the difficulties described in Chapter 4. At any rate the measuring spot of the pattern should be marked, to avoid uncertainties that are given by the unevenness of the pattern. The pattern should be replaced as soon as possible with the first dyeing accepted by the customer (done in many cases in the laboratory) to get a nonmetameric standard of the same material. For production control the standard should be the same as the production. It has to be stored carefully. The measuring spot should, because of the unevenness, be marked. If the same standard is required at several places in the plant, the working standards are tested before using them, so that they will be equal. There are cases described in the literature, where working standards are found, which are different and also metameric, to the standard. Unfortunately this occurs with alarming frequency. When testing colorants the standard always is a sample of the colorant (in general a powder). In that case it has been found that the error of a testing is smallest when the colorant to be tested is dyed or prepared at the same time under the same standardized conditions as the standard for the colorant.

The kind of errors that persist even when we work under optimized conditions shall be discussed with a few examples.

An error caused by the preparation of the sample is the measurement of a sample that is not opaque. For the discussed example thin textile samples are measured. The same spot on each of two yellow, two red, and two blue samples, which have a fairly large color difference, is always measured. The measurements are done with a different number of layers over a black and a white background. In Figure 6-1 the change of hue and chroma of the reference as a function of the number of layers (of the thickness) and of the background is shown in a^*, b^* color diagram. In Tables 6-1–6-3 the color differences between the pairs of samples for the different conditions of measurement are given. Together with the overall color difference, the differences in chroma and hue and the color strength are given. The upper two blocks of numbers (Tables 6-1, 6-2, and 6-3) show the color difference between the two samples as a function of measuring conditions. We see clearly how the color difference changes with the thickness of the sample. We see further that the change is smaller for the measurements over the white background than for that over the black backgrounds. For that reason the sample holders of the instruments should always be white. The measurement error of a nonopaque sample is smaller in

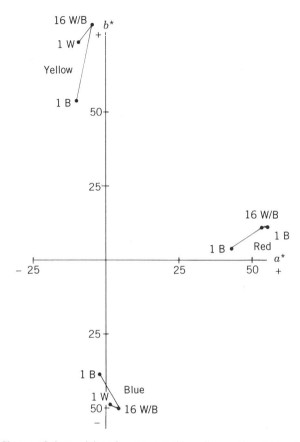

Figure 6-1. Change of chromaticity of transparent thin yellow, red, and blue textile samples as a function of the thickness and the background.

that case. We see that for our example that the sample is opaque only when we measure at least 16 layers. If we look carefully at the numbers, we see a small scatter that is not insignificant. We cannot avoid this when we measure too many layers, because so many layers never give a smooth sample for measuring. (Also drawdowns made with several layers generally do not show a smooth surface.) The data in the three lower blocks in (Tables 6-1–6-3) show the color differences when one sample is measured under different conditions. Especially the two lower blocks of data show how much the color is changed when the sample is measured under the same conditions once not opaque and once opaque. This can be avoided for thin textiles and for paper by choosing the right number of layers although the accuracy may be a little worse. For prints on paper for transfer printing but also for direct prints on paper or textile, the adjustment of the thickness is very difficult. The color difference measured often is caused by the differences in the thickness of the samples to be compared.

Table 6-1. Color Differences between a Pair of Yellow Samples[a]

Yellow		ΔE^* (CIELAB)	ΔC^*	ΔH^*	F (%)
Standard/sample	1 W	2.7	−2.2	−0.1	99
	2 W	4.9	−4.0	0.2	101
	4 W	6.6	−5.0	1.0	100
	8 W	7.4	−6.1	1.4	101
	16 W	7.7	−6.2	2.0	100
	1 B	1.2	0.0	−0.1	100
	2 B	2.3	−2.0	0.3	99
	4 B	4.3	−3.5	0.3	102
	8 B	6.8	−5.6	1.2	102
	16 B	7.3	−6.0	2.0	100
Standard	1 W/1 B	26.0	−19.7	4.1	123
	2 W/2 B	14.8	−11.9	3.6	102
	4 W/4 B	5.7	−4.4	2.4	100
	8 W/8 B	0.8	−0.5	0.6	98
	16 W/16 B	0.4	−0.1	0.2	98
	16 W/8 W	0.6	−0.2	0.5	101
	16 W/4 W	1.2	−0.3	1.1	99
	16 W/2 W	2.6	−0.9	2.4	99
	16 W/1 W	7.3	−5.5	4.8	81
	16 B/8 B	1.5	−0.9	1.1	100
	16 B/4 B	6.2	−4.5	3.2	101
	16 B/2 B	16.0	−12.4	5.6	103
	16 B/1 B	30.9	−24.8	8.1	104

[a]Measured with different number of layers over black and white backgrounds; also color differences between the yellow standard measured with different number of layers over black and white backgrounds. (1 W = one layer over white; F = color strength).

In the next example the results of the individual measurements on a yarn sample are given. The sample is uneven although dyed under optimized conditions. Table 6-4 shows the results measured on a yarn wound on a card. Because a dyeing with only one dyestuff is measured, the different spots show only differences in color strength. The maximum color difference between the measured places is 0.4 CMC (2 : 1) units. It should also be noted that the color strength calculated each time against the average differs more between standard and sample than within one sample. Statistically we must expect a factor of 1.4 for the scattering. In the literature the coefficient of variation for the uniformity of a laboratory dyeing with only one colorant is 0.5–1.0%. This means that different spots of a dyeing may show differences in color strength of 2–4%. Because these numbers are determined during the testing of colorants, where substrate and dyeing condition are optimized, we can expect that production batches where large amounts of material are to be dyed uniformly give a much larger scatter. Because here always more than one colorant is used,

Table 6-2. Color Differences between a Pair of Red Samples[a]

Red		ΔE^* (CIELAB)	ΔC^*	ΔH^*	F (%)
Standard/sample	1 W	2.4	−1.6	−0.7	109
	2 W	4.0	−3.5	−1.5	99
	4 W	4.6	−4.0	−1.8	99
	8 W	4.1	−3.6	−1.8	96
	16 W	3.7	−3.3	−1.5	97
	1 B	1.2	−1.0	−0.5	100
	2 B	2.3	−2.0	−0.9	99
	4 B	3.9	−3.3	−1.4	101
	8 B	4.5	−3.9	−1.6	100
	16 B	3.9	−3.4	−1.9	98
Standard	1 W/1 B	16.4	−13.1	−5.3	127
	2 W/2 B	7.5	−6.5	−2.8	99
	4 W/4 B	1.8	−1.7	−0.7	98
	8 W/8 B	0.7	0.1	−0.1	95
	16 W/16 B	0.2	0.0	−0.1	98
	16 W/8 W	0.4	0.0	0.2	103
	16 W/4 W	0.7	0.6	0.2	100
	16 W/2 W	2.0	1.2	0.8	101
	16 W/1 W	4.3	0.5	0.1	76
	16 B/8 B	0.3	0.2	0.1	99
	16 S/4 B	1.2	−1.8	−0.4	99
	16 S/2 B	5.5	−4.9	−1.9	102
	16 S/1 B	14.4	−12.6	−5.3	98

[a]Measured with a different number of layers over black and white backgrounds; also color differences between the red standard measured with a different number of layers over black and white background. (1 W = one layer over white; F = color strength).

such dyeings not only show differences in color strength but also in chromaticity (for that see Figure 5.3.3-1). Especially for continuous and leather dyeings the differences in color are often quite evident. In addition to the nonuniformity of areas that are not far from each other, there are differences between both edges and between the beginning and the end of the dyeing for each piece of material (textile and paper). We can see them only when we take samples from different parts of the piece. That is not possible when we take a sample for the correction of the batch, but it is possible for the quality control of the batch. For the correction of the batch we must believe that the sample taken is representative for the whole lot. Unfortunately this frequently is not the case, and leads to additional errors.

The next example (Table 6-5) shows color differences calculated between measurements of patterns at different times. They are the patterns from the textile dyeings from Chapters 2 and 3. The color difference between the two

Table 6-3. Color Differences between a Pair of Blue Samples[a]

Blue		ΔE^* (CIELAB)	ΔC^*	ΔH^*	F (%)
Standard/sample	1 W	1.8	−1.4	−0.4	104
	2 W	2.5	−2.2	−1.0	101
	4 W	2.9	−2.5	−1.5	100
	8 W	2.7	−2.4	−1.3	102
	16 W	2.5	−2.1	−1.1	103
	1 B	1.1	−0.9	−0.5	101
	2 B	1.8	−1.6	−0.7	101
	4 B	2.5	−2.1	−1.3	100
	8 B	2.7	−2.3	−1.3	101
	16 B	2.6	−2.2	−1.3	101
Standard	1 W/1 B	12.6	−10.2	−3.6	132
	2 W/2 B	5.9	−5.0	−2.9	101
	4 W/4 B	1.0	−0.7	−0.7	97
	8 W/8 B	0.6	0.0	−0.3	97
	16 W/16 B	0.4	−0.2	−0.2	98
	16 W/8 W	0.3	0.0	0.0	101
	16 W/4 W	0.5	0.4	0.3	102
	16 W/2 W	1.8	1.6	0.4	98
	16 W/1 W	5.6	−0.8	−3.1	76
	16 B/8 B	0.3	0.1	−0.1	98
	16 B/4 B	0.5	−0.3	−0.3	98
	16 B/2 B	4.1	−3.3	−2.4	99
	16 B/1 B	12.9	−11.0	−6.3	100

[a]Measured with a different number of layers over black and white backgrounds; also color differences between the blue standard measured with a different number of layers over black and white backgrounds (1 W = one layer over white; F = color strength).

Table 6-4. Differences in Color Strength (against the Average, in %) between Different Spots on Two Yellow Yarns Wound on Cards and between both Samples, When the Color Strength is Calculated with the Measurement of One Spot Each

Spot	Standard	Sample	Sample/Standard
1	−0.3	+1.8	+2.1
2	+1.0	−1.2	−2.2
3	+0.9	−1.5	−2.4
4	+0.5	+0.6	0.0
5	0.0	+1.3	+1.3
6	−2.1	−1.1	+1.0

Table 6-5. Color Differences between the Measured Results of Samples to Be Matched, Measured at Different Times

Sample	ΔE CMC $(2:1)$
1 (yarn)	1.5
2	0.2
8	0.2
9	0.5
10	0.5
11 (yarn)	0.4
13	0.4
14	0.3
15	0.6

measurements sometimes has the same size as the color tolerance for the delivery. The reason for these differences may be due to the calibration of the instrument. As said before for that example, not only the absolute tristimulus values of pattern and delivery but also the color differences between both show considerable scatter. This may be due to the unevenness of the sample to be matched. Another reason may be the nature of the sample. Samples that consist of a small number of thin fibers cannot always be measured reproducibly.

The examples described above show clearly that each sample should be measured at as many places as possible. The average then will show more accuracy. If the program provides the calculation, the average and the amount of scatter should be printed. Such data are shown in sections a and b of Figures 5.2.1-1 and 5.2.1-2 where the weighted K/S value (formula 5.2.1-4) for the measurements of each spot is printed. The differences in color strength are similar to those of Table 6-4. Small patterns should be measured several times. The computer color matching should be done for the average of the measurements.

A sample not only is uneven but it also cannot be produced exactly the same every time. If we dye, at the same time, three theoretically equal samples as accurately as possible and we measure each sample at different spots, the average values will differ. The difference of the average values generally will be larger than the scattering of the values for different spots of one dyeing. The error due to the reproducibility of a dyeing also is printed in section b in Figures 5.2.1-1 and 5.2.1-2. It is called *intervals*. For textile and paper dyeings it is approximately twice as large as the scatter within a sample. Data for the accuracy of textile dyeings can be found in the report of the ISCC Committee 22 (ISCC[2]). The chairman of the committee at that time, Mr. R. Kuehni, later published a paper with newer results. They are similar to the data published by the committee. The results are confirmed by the producers of colorants. For quality control of colorants therefore generally several (two to four) dyeings of standard and batch are dyed and measured at several spots. Only then will the results have the required accuracy.

The scatter of the results get larger when dyeings, done under carefully chosen conditions, are dyed at different (8–20) times, as is shown in Table 6-6 (Kuehni[2]). For one time, the average of three dyeings done at the same time is used for the calculations. All dyeings are done with the same standardized material and with the same colorant. That is an ideal case that never will be found in industrial practice. Kuehni's paper describes a lot of factors that influence the accuracy of dyeings.

The examples show that it is difficult to produce samples reproducibly. That is true also when we work as accurately as possible. The accuracy can be improved only when we dye in the laboratory every time more than one sample and when we take more than one sample from a production batch. Each sample must be measured at several spots. Because of the speed of modern instruments, the measurements take only little time.

To work with as few dyeings or corrections as possible to get a result that is usable, we recommend to test the reproducibility of dyeings in the laboratory as well as in the plant. If we minimize such sources of error, we save a lot of money (and time). The more accurate the dyeing is the better is, for instance, the result of a formula calculated by computer color matching and the result of a correction. The more accurate the sample taken from a production lot is, the faster it can be corrected. The more accurate the result of quality control is, the fewer discussions there will be between supplier and customer, when both do the test with the same accuracy.

If the result of a measurement does not come up to our expectation, we recommend that the color measurement not be blamed for it—because there are also limits to the accuracy of color measurement—but to determine whether the sample is really representative.

Table 6-6. Differences in Color Strength of Dyeings[a]

Repeatability of Dyeings	Strength (COV %)
A. Single-specie disperse dye on PES ($N = 19$)	1.15
B. Basic dye on acrylic ($N = 8$)	1.53
C. Basic dye on acrylic ($N = 11$)	1.87
D. Multiple-species disperse dye on PES ($N = 20$)	1.85
E. Acid dye on nylon ($N = 18$)	2.68
F. Reactive Red on cotton ($N = 10$)	3.50
G. Reactive Blue on cotton ($N = 20$)	3.63
H. Vat dye on cotton ($N = 12$)	5.14

[a]Done at different times with standardized conditions (substrate, dyestuff, conditions of dyeing); the coefficient of variation (COV) is given; the maximum differences may be four times as large (N = number of dyeings used for calculating the coefficient of variation) (from Kuehni[2]).

Summary. Samples tested with color measurement are for many reasons not representative for the whole lot of the product to be tested. Often the lot is not homogenous and does not have the same color throughout. Also the sample taken for the test or the sample prepared in the laboratory is more or less nonuniform.

In the laboratory it is not possible to prepare samples absolutely reproducibly. The result of the measurement therefore is more accurate when more samples are taken or produced and more places on each sample are measured.

When a color measurement gives an unexpected result, often not the color measuring instrument or the theory, but a nonrepresentative sample is the reason.

Appendix

SYMBOLS AND TERMS

These are listed in the order of their appearance and description in the text.

λ. Symbol for the wavelength. Unit meter (m), statement in nanometers (1 nm $= 1 \times 10^{-9}$ m). The visible wavelength range goes from ~ 380 nm to ~ 780 nm. For color measurement the wavelength range often is reduced to 400–700 nm. The results of the color measurement of fluorescent samples are influenced by the ultraviolet (UV) wavelength region of the spectral power of the light, which ranges from 300 to ~ 380 nm.

   If (λ) (in parentheses) appears after other symbols, it means that they are wavelength-dependent.

$S(\lambda)$. Relative spectral power of light. Unit 1. Wavelength range 300 to ~ 800 nm. For color measurement we do not work with the radiant power $P(\lambda)$ (unit W = watt) of a light source, but with relative spectral power $S(\lambda)$. Often for $S(\lambda)$ the radiation of the wavelength 560 nm is set at 100. The CIE has recommended, for work with color measurement, only a few relative spectral power distributions (standard illuminants). They are described in detail in CIE Publication 15.2, *Colorimetry*.

Standard illuminant A. Standard illuminant A is the relative spectral power distribution of a Planckian radiator (black body) with a color temperature of 2856K. It is similar to that of a tungsten-filament lamp.

Standard illuminant C. Standard illuminant C corresponds to the relative spectral power of daylight with a color temperature of about 6800K. It is defined only from 320 nm. The UV content of standard illuminant C do not correspond with that of daylight. Standard illuminant C is replaced by standard illuminant D65 and should not be used anymore.

Standard illuminant D65. By the CIE this is written D_{65}. Standard illuminant D65 shows the relative spectral power distribution of daylight with a color temperature of 6500K. The spectral power distribution of daylight was measured in several laboratories over a long period. Standard illuminant D65 should be used to describe the color of a sample when illuminated by daylight.

Standard illuminants D50, D55, and D75. These are written D_{50}, D_{55}, and D_{75} by the CIE. The relative spectral power distribution of these standard illuminants is measured in the same manner as that of standard illuminant D65. D50, D55, and D75 should be used only if necessary. In the United States D75 is used in many color matching booths.

Illuminants F1 through F12. These illuminants show the relative spectral power distribution of 12 different fluorescent lamps from 380 to 780 nm. As yet they are not officially recommended, that is, standardized. The use of illuminants F2 (correlated color temperature 4230K), F7 (correlated color temperature 6500K), or F11 (correlated color temperature 4000K) is preferred.

Illuminants XE. These illuminants describe the relative spectral power distribution of xenon high-pressure or xenon flash lamps. Xenon lamps have a relative spectral power distribution that is most similar to that of daylight (also in the UV range). They are therefore often used in color measurement instruments.

$R(\lambda)$. Reflectance factor. Unit: %. The reflectance factor describes the amount of light reflected from the sample in comparison to the amount of light reflected from a perfect diffuser. The reflectance factor is used here as a collective term. The CIE recommends, depending on the conditions of color measurement, the use of different words and different symbols to describe the reflected light—reflectance factor, radiance factor, reflectometer value, and so on. Because the understanding of the different terms is unimportant for color measurement, the terms are not discussed. In the standards the reflectance factor ranges between 0 and 1. In practice the term is modified by multiplying it by 100 and describing the reflectance factor in percent.

$T(\lambda)$. Transmittance factor. Unit: %. The transmittance factor describes the amount of light transmitted through the sample in comparison to the amount of light that falls on the sample. The transmittance factor is used here as a collective term. The CIE recommends, depending on the conditions of color measurement, the use of different words and symbols to describe the transmitted light—transmittance factor, overall transmittance, and so on. Because the understanding of the different terms is unimportant for color measurement, the terms are not discussed. In the standards the transmittance factor has values between 0 and 1. In practice the term is modified by multiplying it by 100 and describing the reflectance factor in percent.

2° standard observer. In the language of the CIE: CIE 1931 standard colorimetric observer.

10° standard observer. In the language of the CIE: CIE 1964 supplementary standard colorimetric observer.

$\bar{x}(\lambda)$, $\bar{y}(\lambda)$, $\bar{z}(\lambda)$. CIE color matching functions for the 2° standard observer.

$\bar{x}_{10}(\lambda)$, $\bar{y}_{10}(\lambda)$, $\bar{z}_{10}(\lambda)$. CIE color matching functions for the 10° standard observer.

$S(\lambda) \times \bar{x}(\lambda)$, $S(\lambda) \times \bar{y}(\lambda)$, $S(\lambda) \times \bar{z}(\lambda)$. Tristimulus weighting factors. They are calculated for many illuminant–observer conditions and for different wavelength intervals. In modern color measurement instruments, several of them are always stored. The most complete collection is given in ASTM E 308-85.[1]

X, Y, Z. Tristimulus values for the 2° standard observer.

X_{10}, Y_{10}, Z_{10}. Tristimulus values for the 10° standard observer.

The statement of tristimulus values as described above is not definite, because it does not say with which illuminant they are calculated. The statement of the illuminant that is used for the calculation is absolutely necessary. The style described below without using the subscript is often used today:

$$X \text{ illuminant–observer, e.g., } X \text{ A/2 or } X \text{ D65/10}$$

Similar statements for Y and Z are necessary.

X_n, Y_n, Z_n. Tristimulus values of a perfect diffuser, or for absolute white. For statement see: **X, Y, Z** tristimulus values.

x, y, z. chromaticity coordinates. For statement see: **X, Y, Z** tristimulus values.

ΔE. Color difference.

ΔE^*. Color difference calculated with a standardized formula.

For the statement of a color difference the formula with which it is calculated has to be given. A statement of the illuminant–observer condition with which the tristimulus values are calculated is also necessary. It is recommended, besides indicating the color difference, to show the components of the color difference (chromaticity and lightness difference; hue, chroma, and lightness difference).

a^*, b^*, L^*. Coordinates of the CIELAB system: L^* = lightness; a^* = red-green coordinate; b^* = yellow-blue coordinate. a^*, b^*, L^* are calculated from tristimulus values. As with all cases where tristimulus values are used, the illuminant–observer conditions must be specified.

MI. Metamerism index. The statement of a metamerism index needs the statement of the illuminant-observer condition for which the pair of samples match and also the statement of the illuminant-observer condition for which the metamerism index is calculated.

W. Whiteness. If a specific whiteness is cited, the formula used must be specified. The whiteness formula recommended by the CIE should be used.

c. Colorant concentration. Units: %, parts, g/liter, g/kg. For color matching the unit of the calibration dyeings must be the same as for the recipes calculated.

$E(\lambda)$ Extinction. Extinction is a measure for the absorption of light in a transparent sample. In the United States normally the letter $A(\lambda)$ (absorption) is used.

$a(\lambda)$ Absorption constant in the Lambert–Beer law. For the calculation of $a(\lambda)$ the thickness of the sample often is included.

F. Color strength. Unit: %. The color strength shows the degree of difference in absorption of a sample compared with that of a standard. The color strength of the standard is 100%.

1/F. Dyeing equivalent. The dyeing equivalent shows how many parts of a colorant are needed, to get a dyeing that looks like a dyeing with 100 parts of the standard.

$K(\lambda)$. Absorption constant in the Kubelka–Munk equation.

$S(\lambda)$. Scattering constant in the Kubelka–Munk equation.

FORMULAS

These are given in the order of their appearance and description in the text.

Calculation of the CIE tristimulus values X, Y, Z (page 27).

$$X = \sum_\lambda S(\lambda)R(\lambda)\bar{x}(\lambda)$$

$$Y = \sum_\lambda S(\lambda)R(\lambda)\bar{y}(\lambda)$$

$$Z = \sum_\lambda S(\lambda)R(\lambda)\bar{z}(\lambda)$$

where
$S(\lambda)$ = relative spectral power, normally a CIE standard illuminant
$R(\lambda)$ = reflectance factor
$\bar{x}(\lambda), \bar{y}(\lambda), \bar{z}(\lambda)$ = 2° or 10° CIE standard observer
λ = wavelength

The statement of tristimulus values must always contain the statement of the illuminant and the observer, which are used for the calculation.

To be absolutely clear, the weighting factors used for the calculation also must be given (step width, range of wavelength). The weighting factors are normalized so that Y_n (Y value for the perfect diffuser) is 100 for every illuminant–observer condition.

Calculation of the CIE chromaticity coordinates (page 31).

$$x = \frac{X}{X + Y + Z}$$

$$y = \frac{Y}{X + Y + Z}$$

The statement of chromaticity coordinates must always contain the statement of the illuminant and the observer, which are used for the calculation.

FMC-2 (Friele, MacAdam, Chickering) color difference formula (page 37).

$$\Delta E = [(\Delta L)^2 + (\Delta C_{r-g})^2 + (\Delta C_{y-b})^2]^{1/2}$$

where $\quad L =$ lightness

$C_{r-g}, C_{y-b} =$ chroma coordinates in the FMC color space

$$\Delta L = \frac{K_2 l}{a}\left[\frac{P\,\Delta P + Q\,\Delta Q}{(P^2 + Q^2)^{1/2}}\right]$$

$$\Delta C_{r-g} = \frac{K_1}{a}\left[\frac{Q\,\Delta P - P\,\Delta Q}{(P^2 + Q^2)^{1/2}}\right]$$

$$\Delta C_{y-b} = \frac{K_1}{b}\left[\frac{S(P\,\Delta P + Q\,\Delta Q)}{P^2 + Q^2} - S\right]$$

where $P = 0.724X + 0.382Y - 0.098Z$

$Q = -0.480X + 1.370Y + 0.1276Z$

$S = 0.686Z$

$$a^2 = \frac{\alpha^2(P^2 + Q^2)}{1 + [NP^2Q^2/(P^4 + Q^4)]}$$

$$b^2 = \beta^2[S^2 + (pY)^2]$$

where $\alpha = 0.00416$

$\beta = 0.0176$

$p = 0.4489$

$N = 2.73$

$l = 0.279$

$K_1 = 0.55669 + 0.049434Y$

$-0.82575 \times 10^{-3}Y^2$

$+0.79172 \times 10^{-5}Y^3$

$-0.30087 \times 10^{-7}Y^4$

$K_2 = 0.17548 + 0.027556Y$

$-0.57262 \times 10^{-3}Y^2$

$+ 0.63893 \times 10^{-5}Y^3$

$-0.26731 \times 10^{-7}Y^4$

or alternatively, according to Judd

$$K_1 = 0.054 + 0.46Y^{1/3}$$

$$K_2 = 0.465K_1 - 0.062$$

CIELAB color difference formula (page 39).

$$\Delta E = [(\Delta a^*)^2 + (\Delta b^*)^2 + (\Delta L^*)^2]^{1/2}$$

or

$$\Delta E = [(\Delta L^*)^2 + (\Delta C^*)^2 + (\Delta H^*)^2]^{1/2}$$

where L^* = lightness
   a^*, b^* = chroma coordinates in the CIELAB color space

$$L^* = 116(Y/Y_n)^{1/3} - 16$$

$$a^* = 500[(X/X_n)^{1/3} - (Y/Y_n)^{1/3}]$$

$$b^* = 200[(Y/Y_n)^{1/3} - (Z/Z_n)^{1/3}]$$

$$C^* = (a^{*2} + b^{*2})^{1/2}$$

$$h = \arctan(b^*/a^*)$$

The formulas apply only when each of the quotients X/X_n, Y/Y_b, and Z/Z_n is greater than 0.008856, which is almost always the case with yarns, skeins, and woven or knitted substrates. If the quotient is smaller than 0.008856, instead of the expression

$$(\text{Quotient})^{1/3}$$

the expression

$$7.787(\text{quotient}) + 16/116$$

should be used.

X_n, Y_n, and Z_n are the tristimulus values for ideal white and the illuminant–observer combination used at each time.

$$\Delta a^* = a_P^* - a_B^* \quad \begin{array}{l} \text{positive} = \text{redder} \\ \text{negative} = \text{greener} \end{array}$$

$$\Delta b^* = b_P^* - b_B^* \quad \begin{array}{l} \text{positive} = \text{yellower} \\ \text{negative} = \text{bluer} \end{array}$$

$$\Delta L^* = L_P^* - L_B^* \quad \begin{array}{l} \text{positive} = \text{lighter} \\ \text{positive} = \text{darker} \end{array}$$

where subscripts P and B represent sample and reference or standard, respectively.

Just as the chromatic colors are described better by their saturation C^* and their hue h with a^* and b^*, the splitting up of the color difference into these coordinates is clearer for such colors.

$$\Delta C^* = C_P^* - C_B^*$$

$$\Delta H^* = [(\Delta E^*)^2 - (\Delta C^*)^2 - (\Delta L)^2]^{1/2}$$

CMC ($l:c$) color difference formula (page 42).

$$\Delta E = \left[\left(\frac{\Delta L^*}{lS_L} \right)^2 + \left(\frac{\Delta C^*}{cS_c} \right)^2 + \left(\frac{\Delta H^*}{S_H} \right)^2 \right]^{1/2}$$

$$S_L = \frac{0.040975 L_1^*}{1 + 0.01765 L_1^*}$$

is $L_1^* < 16$ then

$$S_L = 0.511$$

$$S_c = \frac{0.0638 C_1^*}{1 + 0.0131 C_1^*} + 0.638$$

$$S_H = S_C (Tf + 1 - f)$$

$$f = \left[\frac{(C_1^*)^4}{(C_1^*)^4 + 1900} \right]^{1/2}$$

$$T = 0.36 + |0.4 \cos (h_1 + 35)|$$

has h_1 values between $164°$ and $345°$ then is

$$T = 0.56 + |0.2 \cos (h_1 + 168)|$$

where L_1^*, C_1^*, and h_1 are the values of the standard, l and c are correction factors, to be chosen so that the color difference values can be brought in agreement with visual matchings for certain samples or certain matching conditions.

The vertical lines ($|$ and $|$) enclosing some of the expressions indicate that, for these, the value is always to be taken as positive whatever the numerical result obtained by the initial calculation.

Multiplicative correction for the calculation of corrected metamerism indices (p. 71) is done as follows. For the illuminant–observer condition for which the pair of samples shall match (e.g., D65/10), the tristimulus values

X_{11}, Y_{11}, Z_{11}, for illuminant–observer 1 and sample 1
X_{12}, Y_{12}, Z_{12}, for illuminant–observer 1 and sample 2

are measured and the color difference ΔE_1 is calculated.
The correction factors are calculated as follows:

$$f_X = \frac{X_{11}}{X_{12}}, \qquad f_Y = \frac{Y_{11}}{Y_{12}}, \qquad f_Z = \frac{Z_{11}}{Z_{12}}$$

With these factors the tristimulus values of sample 2 are multiplied. The tristimulus values of sample 2 after the correction are

$$f_X \times X_{12} = X_{11}, \qquad f_Y \times Y_{12} = Y_{11}, \qquad f_Z \times Z_{12} = Z_{11}$$

The color difference between both samples is zero after the correction.
For another illuminant–observer condition the following tristimulus values are calculated for the same pair of samples:

X_{21}, Y_{21}, Z_{21} for illuminant–observer 2 and sample 1
X_{22}, Y_{22}, Z_{22} for illuminant–observer 2 and sample 2

the color difference is ΔE_2.
If one corrects the tristimulus values X_{22}, Y_{22}, Z_{22} with the factors of f_X, f_Y, f_Z one gets the tristimulus values

$$f_X \times X_{22} = X_{23}, \qquad f_Y \times Y_{22} = Y_{23}, \qquad f_X \times Z_{22} = Z_{23}$$

The color difference ΔE_3 between X_{21}, Y_{21}, Z_{21} and X_{23}, Y_{23}, Z_{23} is the metamerism index sought.

Additive correction for the calculation of corrected metamerism (p. 71) indices is done as follows. For the illuminant–observer condition for which the pair of samples shall match (e.g., D65/10) the tristimulus values

X_{11}, Y_{11}, Z_{11} for illuminant–observer 1 and sample 1
X_{12}, Y_{12}, Z_{12} for illuminant–observer 1 and sample 2

are measured and the color difference ΔE_1 is calculated.
From the tristimulus values we calculate the a^*, b^*, L^* values:

a_{11}^*, b_{11}^*, L_{11}^* for illuminant–observer 1 and sample 1
a_{12}^*, b_{12}^*, L_{12}^* for illuminant–observer 1 and sample 2

The correction factors in that case are

$$f_a = a_{11}^* - a_{12}^*, \qquad f_b = b_{11}^* - b_{12}^*, \qquad f_L = L_{11}^* - L_{12}^*$$

These factors are added to the a_{12}^*, b_{12}^*, L_{12}^* values of sample 2. The a^*, b^*, L^* of sample 2 after the correction are

$$f_a + a_{12}^* = a_{11}^*, \qquad f_b + b_{12}^* = b_{11}^*, \qquad f_L + L_{12}^* = L_{11}^*$$

The color difference between both samples is zero after the correction.

For another illuminant–observer condition the following tristimulus values are calculated for the same pair of samples:

X_{21}, Y_{21}, Z_{21} for illuminant–observer 2 and sample 1
X_{22}, Y_{22}, Z_{22} for illuminant–observer 2 and sample 2

the color difference is ΔE_2.

The corresponding a^*, b^*, L^* values are

a_{21}^*, b_{21}^*, L_{21}^* for illuminant–observer 2 and sample 1
a_{22}^*, b_{22}^*, L_{22}^* for illuminant–observer 2 and sample 2

If one corrects the values a_{22}^*, b_{22}^*, L_{22}^* with the factors f_a, f_b, f_L, one gets the a^*, b^*, L^* values:

$$f_a + a_{22}^* = a_{23}^*, \qquad f_b + b_{22}^* = b_{23}^*, \qquad f_L + L_{22}^* = L_{23}^*$$

The color difference ΔE_3 between a_{21}^*, b_{21}^*, L_{21}^* and a_{23}^*, b_{23}^*, L_{23}^* is the metamerism index sought.

Calculation of the overall reflectance R_G of fluorescent samples (p. 94) when changing the illuminating spectral power is done as follows:

$$R_G(\lambda) = R_N(\lambda) + R_F(\lambda)$$

where $R_G(\lambda)$ = total light reflected from the sample at wavelength λ
$R_N(\lambda)$ = normal reflected light from the sample at wavelength λ
$R_F(\lambda)$ = fluorescent light emitted from the sample at wavelength λ

$$R_F(\lambda) = \frac{NF(\lambda)}{S(\lambda)}$$

where $F(\lambda)$ = amount of fluorescent light referred to $N = 1$

$$N = \sum S(\lambda')[1 - R_N(\lambda')]\left[1 - \frac{R_N(\lambda')[1 - R_0(\lambda')]^2}{R_0(\lambda')[1 - R_N(\lambda')]^2}\right]\lambda' \, \Delta\lambda'$$

where λ' = stimulation wavelength range of fluorescence
          $R_0(\lambda')$ = reflectance factor of the material without fluorescent whiten-
                    ing agents
          $R_N(\lambda')$ = normal reflected light from the sample at wavelength λ'

The reliability of $R_N(\lambda')$ is in the wavelength region where fluorescence and absorption exist not good.

CIE whiteness formula for D65/10 (p. 97):

$$W = Y + 800(0.3138 - x) + 1700(0.3310 - y)$$

tint (shade)

$$T = 900(0.3138 - x) - 650(0.3310 - y)$$

If T is positive, the sample is greenish.
If T is negative, the sample is reddish.

Limits of application:

$$Y > 70, \qquad T < \pm 3$$

Lambert–Beer law and equations related to them for calculation of color strength: (pp. 110 to 113):

$$\log 1/T(\lambda) = E(\lambda) = a(\lambda)cd$$

where T in this law has the values $0 - 1$ and $a(\lambda)$ is the absorption constant of the colorant. When d is included in a, we get

$$\log 1/T(\lambda) = E(\lambda) = a(\lambda) \times c$$

Reversed, the law is given by the following equation:

$$T = 10^{-E} - 10^{-ac}$$

The absorption (extinction E) is the sum of the absorption of the material to be dyed and the absorption of the colorant:

$$E = E_M + E_F = E_M + a \times c$$

where subscripts M and F represent material and colorant, respectively.
   When more than one colorant is used, we get, for example, for three colorants

$$E = E_M + E_F = E_M + a_1 \times c_1 + a_2 \times c_2 + a_3 \times c_3$$

Color strength (F)

$$F = \frac{a_P}{a_B}$$

$$F = \frac{E_P - E_M}{E_B - E_M} = \frac{E_{PF}}{E_{BF}}$$

where subscripts P and B represent sample and standard, respectively.
  If the two colorants are used in different concentrations, we get

$$F = \frac{(E_P - E_M)c_B}{(E_B - E_M)c_P}$$

The Kubelka–Munk equations (pp. 114 to 120) and related equations for determining color strength are as follows:

$$\frac{K}{S} = \frac{(1 - R)^2}{2R}$$

where R has the values 0–1; K is a measure for the absorption of light in the sample, and is dependent on the concentration of the colorant; S is the measure for the scattering of light in the sample. Often S is given by the material to be dyed. When using scattering pigments, S is dependent on their concentration.
  Reversed, we get the following equation:

$$R = 1 + \frac{K}{S} - \left[\left(1 + \frac{K}{S}\right)^2 - 1\right]^{1/2}$$

For a nonscattering colorant we get

$$\frac{K}{S} = \left(\frac{K}{S}\right)_M + \left(\frac{K}{S}\right)_F = \left(\frac{K}{S}\right)_M + \left(\frac{K_F}{S}\right)c$$

For a scattering colorant we get

$$\frac{K}{S} = \frac{K_M + K_F c}{S_M + S_F c}$$

when three colorants are used, we get

$$K_F c = K_1 c_1 + K_2 c_2 + K_3 c_3$$

$$S_F c = S_1 c_1 + S_2 c_2 + S_3 c_3$$

For nonscattering colorants, this gives

$$\frac{K}{S} = \left(\frac{K}{S}\right)_M + \left(\frac{K_1}{S}\right) c_1 + \left(\frac{K_2}{S}\right) c_2 + \left(\frac{K_3}{S}\right) c_3$$

For scattering this gives correspondingly

$$\frac{K}{S} = \frac{K_M + K_1 c_1 + K_2 c_2 + K_3 c_3}{S_M + S_1 c_1 + S_2 c_2 + S_3 c_3}$$

Color strength (F):

$$F = \frac{K_{PF}}{K_{BF}}$$

with

$$K_F/S = \frac{(K/S) - (K/S)_M}{c}$$

we get

$$F = \frac{[(K/S)_P - (K/S)_M] c_B}{[(K/S)_B - (K/S)_M] c_P}$$

Because the samples to be compared often have a small shade difference, use of the following equation is recommended:

$$F = \frac{\sum [(K/S)_P - (K/S)_M][\bar{x}(\lambda) + \bar{y}(\lambda) + \bar{z}(\lambda)] c_B}{\sum [(K/S)_B - (K/S)_M][\bar{x}(\lambda) + \bar{y}(\lambda) + \bar{z}(\lambda)] c_P}$$

Bibliography

1. Textbooks (B) (B*—also recommended for beginners), reference books (R), CIE recommendations, and ASTM test methods and practices.

B* F. W. Billmeyer, Jr., and M. Saltzman, *Principles of Color Technology, 2nd ed.,* Wiley, New York, 1981.

B* A. Brockes, D. Strocka, and A. Berger-Schunn, *Color Measurement in the Textile Industry*, Mobay Corp., Pittsburgh, 1989.

B R. W. G. Hunt, *Measuring Colour*, 2nd ed., Hemel Hempstead, Herts Simon & Schuster International Group, 1991.

B* R. S. Hunter, R. W. Harold, *The Measurement of Appearance*, New York, Wiley, 2nd Edition 1987

B D. B. Judd and G. Wyszecki, *Color in Business, Science and Industry*, 3rd ed., Wiley, New York, 1975.

B H. Loos, *Farbmessung*, Beruf + Schule, Itzehoe, 1989.

B* R. McDonald, Ed., *Colour Physics for Industry*, Society of Dyers and Colourists, Bradford, (UK) 1987.

B K. McLaren, *The Color Science of Dyes and Pigments*, 2nd ed., Adam Hilger, Bristol, 1986.

R. G. Wyszecki and W. S. Stiles, *Color Science*, 2nd ed., Wiley, New York, 1982.

R *Colour Index*, 3rd ed., Society of Dyers and Colourists, Bradford 1971, supplement 1976; also regular supplements from time to time.

CIE publications:

CIE Publication 15.2, *Colorimetry*.

CIE Publication 17.4, *International Lighting Vocabulary*.

CIE Publication 80, *Specific Metamerisim Index: Change in Observer, 1989.*

CIE publications are obtainable from the national committees of the CIE. In the United States: U.S. National Committee, USNC/CIE, Gaithersburg MD 20899.

R ASTM test methods for

E 308-85 *Computing the Color of Objects by Using the CIE System*
E 97-82   *Directional Reflectance Factor, 45-deg., 0-deg.*
E 1247-88 *Identifying Fluorescence in Object-Color Specimens by Spectrophotometry*

E 1348-90   *Transmittance and Color by Spectrophotometry Using Hemispherical Geometry*

E 1331-90   *Reflectance Factor and Color by Spectrophotometry Using Hemispherical Geometry*

R ASTM *Practices for*

E 911-90   *Color Measurement of Fluorescent Specimens*

E 805-91   *Identification of Instrumental Methods of Color or Color-Difference Measurement of Materials*

E 1164-91   *Obtaining Spectrophotometric Data for Object-Color Evaluations*

E 259-66   *Preparation of Reference White Reflectance Standards*

D 3134-89   *Selecting and Defining Color and Gloss Tolerances of Opaque Materials and for Evaluating Conformance*

E 1345-90   *Reducing the Variability of Color Measurements by the Use of Multiple Measurements*

E 179-90   *Guide for Selection of Geometric Conditions for Measurement of Reflection and Transmission Properties of Material*

The German standards which are based on the CIE recommendations, are (in Germany: DNK der CIE, Burggrafenstr.2-10, 10787 Berlin),

DIN 5033 Blatt 1-9 Farbmessung

DIN 6172 Metamerieindex von Probenpaaren bei Lichtwechsel

DIN 6173 Blatt 1-2 Farbabmusterung

DIN 6174 Farbmetrische Bestimmung von Farbabständen bei Körperfarben nach der CIELAB-Formel

2. Papers and books cited within this book (in alphabetic order of the authors).

W. Baumann et al., "Determination of Relative Colour Strength and Residual Colour Difference by Reflectance Measurement," *J. Soc. Dyers Colourists*, **103**, 100–105 (1987).

A. Berger, Visuelle Farbabmusterung, *Die Farbe*, **25**, 33–47 (1976).

R. Brossmann et al., "Determination of Relative Colour Strength in Solutions," *J. Soc. Dyers Colourists*, **103**, 38–42 (1987).

CIE Publication 80, *Special Metamerism Index: Changes in Observer*, 1989.

T. R. Commerford, "Difficulties in Preparing Dye Solutions for Accurate Strength Measurements," *Text. Chem. Color.*, **6**, 14–21 (1974).

P. Dorsch, H. Wilsing, and K.-H. Peters, "Die Farbtiefe beeinflussende Parameter, untersucht an Färbungen auf Polyacrylnitril-Fasern," *Melliand Textilberichte*, **62**, 188–193 (1981).

E. Ganz, "Whiteness: Photometric Specification and Colorimetric Evaluation," *Appl. Optics*, **15**, 2039–2058 (1976).

D. Gundlach and E. Mallwitz, "Fragen der Probenbeleuchtung und der Messgeometrie in der Farbmessung," *Die Farbe*, **25**, 113–130 (1976).

ISCC, "A General Procedure for the Determination of Relative Dye Strength by Spectrophotometric Measurement of Reflectance Factor," *Text. Chem. Color.*, **6**, 104–108 (1974).

R. Johnston, "Pitfalls in Color Specification," *Off. Dig.*, **35**, 259–274 (1963).

R. Kuehni, "Repeatability of Dyeing," *Text Chem. Color.*, **21**, (8), 23–25 (1988).

M. R. Luo and B. Rigg, "BFD(l : c) Colour Difference Formula Part 2–Performance of the Formula," *J. Soc. Dyers Colourists*, **103**, 126–132 (1987).

D. MacAdam, "Maximum Visual Efficiency of Colored Materials," *J. Opt. Soc. Am.*, **25**, 361–367 (1935).

S. Rösch, "Darstellung der Farbenlehre für die Zwecke des Mineralogen," *Fortsch. Mineral. Kristallogr. Petrogr.*, **13**, 73–234 (1929).

E. I. Stearns, *The Practice of Absorption Spectrophotometry*, Wiley, New York, 1969.

D. Strocka, "Are Intervals of 20 nm Sufficient for Industrial Colour Measurement?" *COLOUR 73*, Adam Hilger, London, 1973, 453–456.

W. D. Wright, *The Measurement of Color*, Adam Hilger, London, 4th Edition 1969.

G. Wyszecki, "Development of New Standard Sources for Colorimetry," *Die Farbe*, **19**, 43–76 (1970).

3. Some of the large and important producers of color measurement systems, color-imeters, artificial daylighting units, and color standards are listed below. The major manufacturers of colorants, in general, have up-to-date knowledge of the developments in this field. Before buying a system, one should get information from them.

Three of the larger producers of color measurement systems for color matching and quality control have merged into one company. The new company is called datacolor international:

in USA:
3735 Beam Road, Charlotte, NC 28217 (USA)
Tel. (1) 704 357 0400

in United Kingdom:
6 St. George's Court, Dairyhouse Lane,
Broadheath, Altrincham, Cheshire WA14 5UA (UK)
Tel. (44) 61 929 9441

in Germany:
In den Seewiesen 68, 89520 Heidenheim (D)
Tel. (49) 7321 35980

The following companies sell color measurement systems for color matching. They also sell special instruments and visual color matching units.

Hunter Associates Laboratory, Inc.
11495 Sunset Hills Road, Reston, VA 22090-5280 (USA)
Tel. (1) 703 471 6870
An important producer of colorimeters and gloss meters, also of instruments for special purposes.

Macbeth, Division of Kollmorgen
Little Britain Road, P.O. Box 230, Newburgh, NY 12550 (USA)
Tel. (1) 800 491 4952
Also an important producer of visual matching units.

E 1349-90 *Reflectance Factor and Color Using Bidirectional Geometry*
E 284-91b *Standard Terminology of Appearance*

British Ceramic Research Assoc.
Queens Road, Penkhull, Stocke-on-Trent, (UK)
Sells 12 ceramic standards. They are sold standardized by the National Physical Laboratory (NPL) or not standardized. Available in the USA from Dr. H. Hemmindinger, 438 Wendover Drive, Princeton, NJ 8540.

Collaborative Testing Services Inc.
PO Box 1049, Herndon, VA 22070 (USA)
Provides samples to test the accuracy of color measuring instruments.